Lecture Notes
in Control and Information Sciences 188

Editors: M. Thoma and W. Wyner

D. Subbaram Naidu

Aeroassisted Orbital Transfer

Guidance and Control Strategies

Springer-Verlag London Ltd.

Series Advisory Board

Author

D. Subbaram Naidu, PhD
Measurement and Control Research Center, College of Engineering,
Idaho State University, Pocatello, Idaho 83209, USA

ISBN 978-3-540-19819-2   ISBN 978-3-540-39309-2 (eBook)
DOI 10.1007/978-3-540-39309-2

Typesetting: Camera ready by author

69/3830-543210 Printed on acid-free paper

Dedicated whole heartedly to

Douglas B. Price

without whose interest, support, cooperation, and understanding

this monograph certainly would not have been possible!

"Man must rise above the Earth...to the top of the atmosphere and beyond...for only thus will he understand the world in which he lives."

-Socrates, 400 B.C.

"I believe that this nation should commit itself to achieving the goal, before this decade is out, of landing a man on the Moon and return him safely to the Earth".

-John F. Kennedy, May 25, 1961

"And then...a journey into tomorrow...a journey to another planet...a manned mission to Mars."

-George Bush, July 20, 1989

PREFACE

According to the report of the National Commission on Space, PIONEERING THE SPACE FRONTIER, the concept of aerobraking for orbital transfer has been recognized as one of the critical technologies and recommended for demonstration projects in building the necessary technology base for pioneering the space frontier. In space transportation systems, aerobraking (or aeroassist), defined as the deceleration resulting from the effects of atmospheric drag upon a vehicle during orbital operations, opens new mission opportunities, especially with regard to the establishment of a permanent space station and space explorations to other planets such as Mars. The main function of a space transportation system is to deliver payloads from Earth to various locations in space. Until now, this function has been performed by various rockets, the space shuttle, and expendable upper stages using solid or liquid propellants. In particular, considering the economic benefits and reusability, an orbital transfer vehicle (OTV) is proposed for transporting payloads between low Earth orbit (LEO) and high Earth orbit (HEO). The two basic operating modes contemplated for OTV are a ground-based OTV which returns to Earth after each mission and a space-based OTV which operates out of an orbiting hanger located at the proposed Space Station Freedom.

The main areas of research that are reported in this monograph are atmospheric entry problem, orbital transfer with aeroassist technology, aerocruise, and guidance. Using the theory of optimal control, and the method of matched asymptotic expansions, Chapter 1 presents the atmospheric entry problem. In Chapters 3, and 4, using algorithms based on the industry standard program to optimize simulated trajectories (POST), and the multiple shooting method, we describe methods to generate fuel-optimal trajectories for planar orbital transfer and noncoplanar orbital transfer. Chapter 5 consideres cruise maneuver being performed using either bank control with constant thrust, or thrust control with constant bank control and obtain conditions for maximum plane change for a given fuel consumption. In Chapter 6, guidance schemes for atmospheric maneuver under deterministic conditions are presented. Finally, the monograph ends with a bibliography on this topic to provide the reader with a literature status for further research.

I take this opportunity to express my deep gratitude to Dr. Douglas B. Price, Head of Spacecraft Controls Branch, Guidance and Control Division, NASA Langley Research Center, Hampton, Virginia for his support, interest, cooperation, and understanding throughout my stay at NASA Langley, and most appropriately I dedicate this Research Monograph to him. Thanks are due to Dr. Daniel Moerder of NASA Langley for acting as technical monitor and for his understanding and cooperation to carry out my research. The support of my colleague Dr. Joe Hibey and graduate student Mr. C. Charalambous of Old Dominion University, Norfolk, Virgina, is greatly appreciated.

Special thanks go to Dr. Hary Charyulu, Dean of the College of Engineering, Idaho State University, Pocatello, Idaho, who has always been more than willing to create an atmosphere conducive to research activities which led to the final preparation of this monograph. I would like to thank my colleague Dr. Kevin Moore for some useful discussions on guidance and control. The help rendered by Linda, Michelle, and Trina at the office of the College of Engineering deserves special mention. Additionally, my graduate student Mr. Lin Li was particularly helpful in the preparation of some of the figures for the monograph.

I also extend a note of thanks to Dr. Aaron D. Wyner, Editor, Lecture Notes in Control and Information Sciences for his keen interest in publishing this monograph. At Springer-Verlag, Mr. Nicholas Pinfield and Lynda Mangiavacchi who has contributed in many ways to enhance the professional appearance of the monograph deserve many thanks.

Finally, I have been fortunate to have an uninterrupted love, support and understanding from my wife Sita, and two daughters Radhika, and Kiranmai for all my academic activities.

D. Subbaram Naidu
Pocatello, Idaho
June, 1993

CONTENTS

CHAPTER 5. ORBITAL PLANE CHANGE WITH AEROCRUISE

CHAPTER 6. OPTIMAL GUIDANCE FOR ORBITAL TRANSFER

CHAPTER 1

INTRODUCTION

The specification spectrum for the proposed Space Transportation System (STS) places heavy emphasis on the development of reusable avionics subsystems having special features such as vehicle evaluation and reduction of ground support for mission planning, contingency response and verification and validation. According to the report of the National Commission on Space, PIONEERING THE SPACE FRONTIER [1], the concept of aerobraking for orbital transfer has been recognized as one of the critical technologies and recommended for demonstration projects in building the necessary technology base for pioneering the space frontier. In space transportation systems, aerobraking (or aeroassist), defined as the deceleration resulting from the effects of atmospheric drag upon a vehicle during orbital operations, opens new mission opportunities, especially with regard to the establishment of a permanent space station and space explorations to other planets such as Mars [2]. Fig. 1 shows the space transportation architecture relevant to aeroassist technology.

The main function of a space transportation system is to deliver payloads from Earth to various locations in space. Until now, this function has been performed by various rockets, the space shuttle, and expendable upper stages using solid or liquid propellants. In particular, considering the economic benefits and reusability, an orbital transfer vehicle (OTV) is proposed for transporting payloads between low Earth orbit (LEO) and high Earth orbit (HEO). The two basic operating modes contemplated for OTV are a ground-based OTV which returns to Earth after each mission and a space-based OTV which operates out of an orbiting hanger located at the proposed Space Station Freedom.

The main areas of research that are reported in this monograph are atmospheric entry problem, orbital transfer with aeroassist technology, aerocruise, and guidance.

1.1 Atmospheric Entry

The dynamics of many control systems is described by high-order differential equations containing parameters such as small time constants, masses, moments of inertia, inductances, capacitances, Mach number, and high Reynold's number. The presence of these "parasitic" parameters is often the source for the increased order and the "stiffness" of the system. The

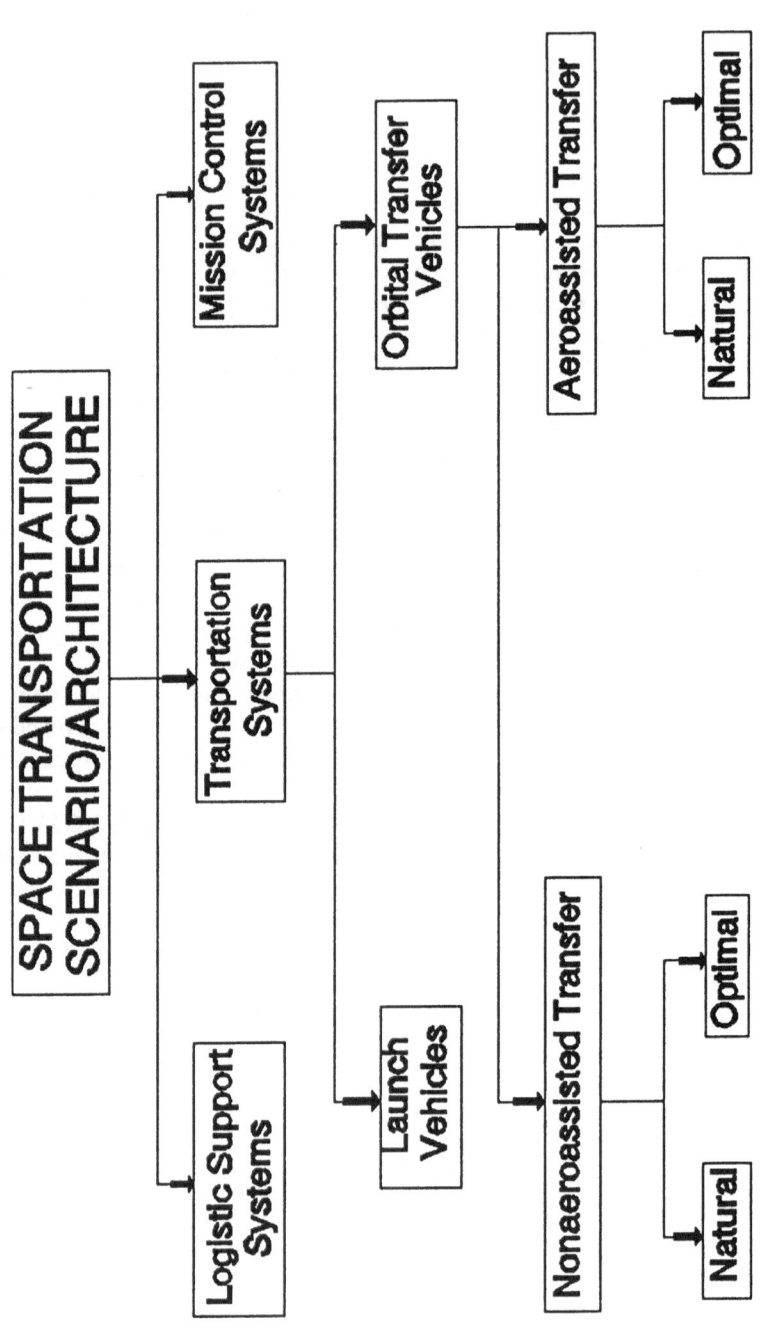

Fig. 1   Space Transportation Architecture

"curse" of the dimensionality coupled with the stiffness poses formidable computational complexities for the analysis and control of such large systems. Singularly perturbed systems are those whose order is reduced when the parasitic parameter is neglected. The methodology of singular perturbations and time scales (SPATS) is a "boon" to control engineers in tackling these large scale systems. As such it is very desirable to formulate many control problems to fit into the framework of the mathematical theory of SPATS.

The methodology of SPATS has an impressive record of applications in a wide spectrum of fields including flight mechanics and trajectory optimization. The aerospace problems involve, in general, the solution of nonlinear differential equations by resorting to numerical integration. Analytical solutions are important in providing a general understanding of the structure of solutions and a better foundation for the solution of guidance problems. With this in view, attempts have been made to obtain approximate analytical solutions for the atmospheric entry problem using asymptotic methods such as the method of matched asymptotic expansions, singular perturbation method, and multiple scale method. In Chapter 2 of this monograph, using the theory of SPATS, the atmospheric entry problem is presented [3-8].

1.2 Aeroassist (Aerobraking) Technology

In a typical mission, a space-based OTV, which is initially at the space station orbit (SSO), is required to transfer a payload to geosynchronous Earth orbit (GEO), pick up another payload, such as a faulty satellite, and return to mate with the orbiting hanger at SSO for refurbishment and redeployment of the payload. The OTV on its return journey from GEO to SSO needs to dissipate some of its orbital energy. This can be accomplished by using an entirely propulsive (Hohmann) transfer in space only or a combination of propulsive transfer in space and aeroassisted maneuver in the atmosphere [Fig. 2]. It has been established that a significant (of the order of 50%) fuel savings and the resulting increased payload capabilities can be achieved with propulsive and aerobraking (or aeroassisted) maneuvers instead of all-propulsive maneuvers [2]. The phrase "aeroassisted" is often used to convey that the atmosphere is used to assist the desired deceleration. This leads to an aeroassisted orbital transfer vehicle (AOTV) which, on its return leg of the mission, dips into the Earth's atmosphere, utilizes atmospheric drag to reduce the orbital velocity and employs lift and bank angle modulations to achieve a desired orbital inclination. Basically, the AOTV performs a synergistic maneuver, employing a hybrid combination of propulsive maneuver in space and aerodynamic maneuver in the atmosphere. Broadly speaking, the two kinds of orbital transfer are coplanar orbital transfer and noncoplanar orbital transfer (or

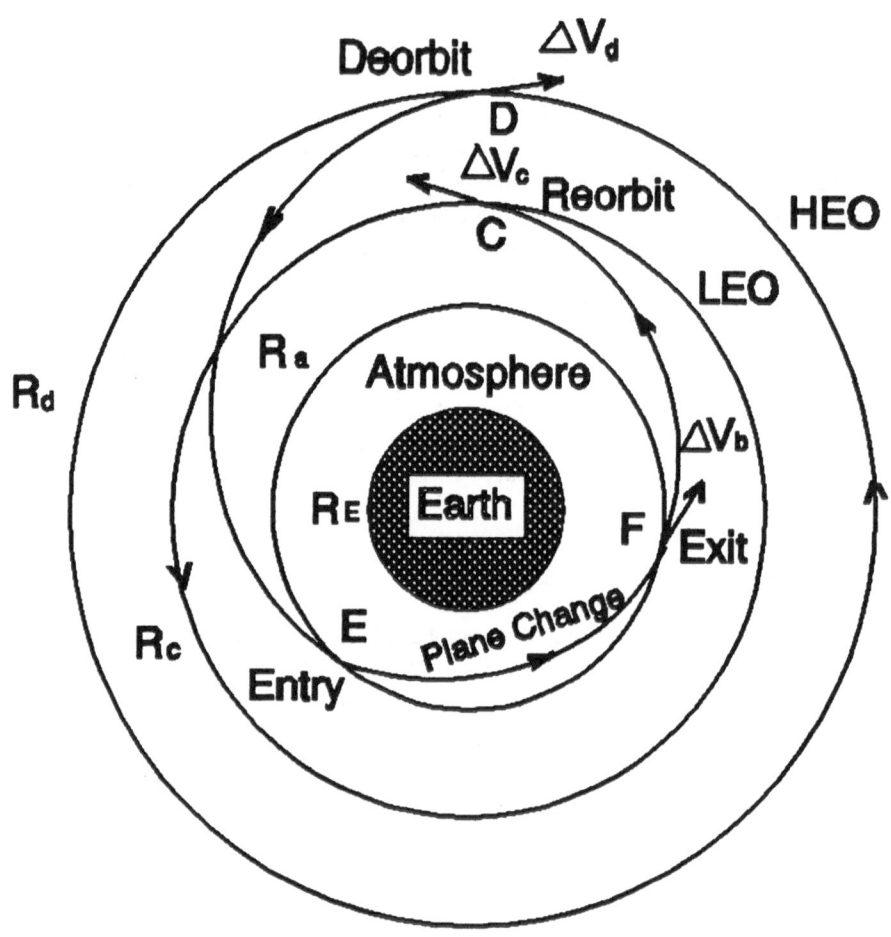

Fig. 2   Aeroassisted Orbital Transfer

orbital transfer with plane change). The fuel optimization is an important aspect of orbital transfer missions.

In Chapters 3, and 4 of this monograph, using algorithms based on the industry standard program to optimize simulated trajectories (POST), and multiple shooting method, are described methods to generate fuel-optimal trajectories for coplanar orbital transfer [9-13], and noncoplanar orbital transfer [14-16]. Chapter 3 briefly describes the various types of coplanar transfers. Then we address the fuel-optimal control problem arising in coplanar orbital transfer employing aeroassist technology. The maneuver involves a transfer from HEO to LEO and at the same time minimization of the fuel consumption for achieving the desired orbit transfer. It is known that a change in velocity, also called the characteristic velocity, is a convenient parameter to measure the fuel consumption. A suitable performance index is the total characteristic velocity which is the sum of the characteristic velocities for deorbit and for reorbit (or circularization). Use of Pontryagin minimum principle leads to a nonlinear, two-point boundary value problem (TPBVP) in state and costate variables. Here, aerodynamic lift is used as the only control variable. The solution of the TPBVP is the stumbling block in obtaining fuel-optimal solution. This problem is solved by using a more efficient multiple shooting method which is a simultaneous application of a single shooting algorithm to equally divided points of the total interval of the solution [11]. The fuel-optimal solution forms as a basis for on-board guidance of an AOTV [9-13].

Chapter 4 discusses the fuel minimization for the more important noncoplanar orbital transfer. Here, a desired plane change is achieved by a combination of aerodynamic lift and bank angle modulations so as to bring the vehicle to the required orbital plane at LEO to rendezvous with the orbiting space station. The maneuver involves the transfer from HEO to LEO with a plane change and at the same time minimization of the fuel consumption. For the minimum-fuel maneuver, the objective is then to minimize the total characteristic velocity for deorbit, boost, and reorbit (or circularization) for a specified change in inclination angle. Application of Pontryagin minimum principle leads to a nonlinear, two-point, boundary value problem (TPBVP), which is solved by using the multiple shooting method [14-16].

1.3 Aerocruise

There are basically three methods of plane change, (i) the impulsive method, (ii) the aeroglide method, and (iii) the aerocruise method. In the impulsive method, the plane change is achieved entirely outside the atmosphere, and fuel consumption is prohibitively large for sizable changes of orbital plane. In both the aeroglide and aerocruise methods, rockets are used to deflect the vehicle into the atmosphere, and the plane change is accomplished by heading change of the vehicle. With aeroglide there is no thrusting during the atmosphere, and with aerocruise, atmospheric drag is balanced by a continuous thrust to keep the spacecraft at a constant altitude and velocity. Propellant expenditure comparisons among the three methods of plane change show that the aerocruise method is superior to other competing methods for plane changes greater than about 20 degrees, and with heating restraints. The basic effect of propulsion during aerocruise is to (i) balance drag in order to maintain constant velocity, (ii) augment lift with a component of thrust, thus increasing cruising altitude over what it would be during aeroglide turn, and finally (iii) provide a lateral component of thrust giving the required turn necessary for plane change.

Chapter 5 considers cruise maneuver being performed using either bank control with constant thrust, or thrust control with constant bank control and obtains conditions for maximum plane change for a given fuel consumption. This investigation is carried out for aerobraking at Earth and Mars. [17-19].

1.4 Guidance

An optimal trajectory is computed for a given nonlinear dynamical system with a fixed set of conditions. However, variation of the initial and final conditions, plant parameters would alter the optimal trajectory. It is computationally tedious and expensive to repeat the whole optimization procedure for every changed condition and obtain a new optimal trajectory. In such a situation, an alternative is to linearize the original system and generate an optimal trajectory in the neighborhood of the original optimal trajectory, involving considerably less computational effort.

Guidance is the determination of a strategy for following a nominal flight in the presence of off-nominal conditions, wind disturbances, and navigation uncertainties. In a typical guidance scheme, the final steering command is generated as the sum of two components, an open-loop actuating (control) signal yielding the desired vehicle trajectory in the absence of external disturbances, and a linear feedback regulating signal which reduces the system sensitivity to unwanted influences on the vehicle.

Chapters 6 presents a guidance scheme for atmospheric maneuver for deterministic case [20-25] and the fuel-optimal control problem arising in noncoplanar orbital transfer employing aeroassist technology. The maneuver involves the transfer from HEO to LEO with a prescribed plane change and at the same time minimization of the fuel consumption. For minimum-fuel maneuver, the objective is to minimize the total characteristic velocity for deorbit, boost, and reorbit (or circularization). The corresponding optimal (nominal) trajectory and control are obtained. The linearization is performed around the nominal condition and the resulting model is fitted into the framework of linear quadratic regulator (LQR) theory. Instead of using time, one of the state variables is employed as an independent variable, thus avoiding any control required due to irrelevant timing errors. Also, the elimination of time as an independent variable carries with it the advantage of order reduction for the system. The choice of weighting matrices in the performance index is made by combining a heuristic method and optimal modal control approach. The feedback control law is obtained to suppress the perturbations from the nominal condition. The results are shown for a typical AOTV [20-22].

Finally, this monograph ends with a bibliography on this topic of guidance and control strategies for aeroassisted orbital transfer in order to provide the reader a literature status for further research.

References

[1] *Pioneering the Space Frontier*, The Report of the National Space Commission on Space, Banton Books Inc., New York, May 1986.

[2] G. D. Walberg, "A survey of aeroassisted orbital transfer," J. Spacecraft, 22, 3-18, 1985.

[3] F. Frostic, and N. X. Vinh, "Optimal aerodynamic control by matched asymptotic expansions," Acta Astronautica,3, 319-332, 1976.

[4] D. S. Naidu and A. K. Rao, Singular Perturbation Analysis of Discrete Control Systems, Lecture Notes in Mathematics, Vol., 1154, Springer Verlag, Berlin, 1985.

[5] D. S. Naidu and D. B. Price, "Time scale synthesis of a closed-loop discrete optimal control system," J. Guidance, Control, and Dynamics, 10, 417-421, 1987.

[6] D. S. Naidu, Singular Perturbation Methodology in Control Systems, IEE Control Engineering Series, Peter Peregrinus Limited, Stevenage Herts, England, 1988.

[7] D. S. Naidu and D. B. Price, Singular perturbations and time scales in the design of digital flight control systems, NASA Technical Paper 2844, Dec., 1988.

[8] D. S. Naidu, "Three-dimensional atmospheric entry problem using method of matched asymptotic expansions," IEEE Trans. on Aerospace and Electronic Systems, 25, 660-667, 1989.

[9] Vinh, N.-X., Optimal Trajectories in Atmospheric Flight, Elsevier

Scientific Publishing Co., Amsterdam, 1981.

[10] K. D. Mease, and N. X. Vinh, "Minimum-fuel aeroassisted coplanar orbit transfer using lift modulation," J. Guidance, Control, and Dynamics, 8, 134-141, 1985.

[11] J. Stoer, and R. Bulirsch, Introduction to Numerical Analysis, Springer-Verlag, New York, 1980.

[12] D. S. Naidu, J. L. Hibey, and C. Charalambous, "Optimal control of aeroassisted coplanar orbital transfer vehicles," 27th IEEE Conference on Decision and Control, Austin, TX, Dec. 7-9, 1988.

[13] D. S. Naidu, J. L. Hibey, and C. Charalambous, "Fuel-optimal trajectories for aeroassisted coplanar orbital transfer problem," IEEE Trans. Aerospace and Electronic Systems, 26, 374-381, 1990.

[14] D. G. Hull, J. M. Glitner, J. L. Speyer, and J. Maper, "Minimum energy loss guidance for aeroassisted orbital plane change," J. of Guidance, Control, and Dynamics, 8, 487-493, 1985.

[15] D. S. Naidu, "Fuel-optimal trajectories for aeroassisted orbital transfer with plane change," AIAA Guidance, Navigation, and Control Conference, Boston, MA, Aug.14-16, 1989.

[16] D. S. Naidu, "Fuel-optimal trajectories for aeroassisted orbital transfer with plane change," IEEE Trans. Aerospace and Electronic Systems, 27, 361-369, 1991.

[17] R. I. Cervisi, "Analytical solutions of a cruising plane change maneuver," J. of Spacecraft & Rockets, 22, 134-140, 1985.

[18] D. S. Naidu, "Orbital plane change maneuver with aerocruise," AIAA 29th Aerospace Sciences Meeting and Exhibit, Reno, Nevada, Jan. 7-10, 1991.

[19] Li, L., and D. S. Naidu, "Orbital plane change maneuver with aerobraking for Mars mission," Research Report, College of Engineering, Idaho State University, Pocatello, Idaho, August 1991.

[20] H. J. Pesch, "Real-time computation of feedback controls for constrained optimal control problems, part I: neighboring extremals," Optimal Control: Applications and Methods, 10, 129-145, 1989.

[21] D. S. Naidu, "Neighboring optimal guidance for aeroassisted noncoplanar

orbital transfer," AIAA Atmospheric Flight Mechanics Conference, New Orleans, LA, Aug. 12-14, 1991.

[22] D. S. Naidu, Guidance and Control Strategies for Aerospace Vehicles, Final Research Report, Dept. of Electrical Engineering, Old Dominion University, Norfolk, Virginia, August 1990.

****

CHAPTER 2

ATMOSPHERIC ENTRY PROBLEM

2.1 Introduction

After accomplishing an assigned mission, a space vehicle enters the atmosphere and uses its aerodynamic maneuverability to reach a prescribed region before the final approach and landing. The entry trajectory, starting at the top of the sensible atmosphere and ending at a low altitude before the final approach and landing, is the most critical portion of the flight path. During this portion, there is an appreciable change in speed, kinetic energy, dynamic pressure and heating rate. It is then of paramount importance to have analytical solutions to the behavior of the trajectory.

The atmospheric entry problem involves, in general, the solution of nonlinear differential equations by resorting to numerical integration. Analytical solutions of a simplified entry problem are important from the point of view of serving as a basis for investigating more complicated cases and providing a general understanding of the structure of solutions. Analytical solutions also provide a better foundation for the solution of guidance problems. With this in view, attempts have been made to obtain approximate analytical solutions for the entry problem using asymptotic methods such as the method of matched asymptotic expansions, singular perturbation method, and multiple scale method [1-11]. Most of these solutions were obtained either for the two-dimensional case, or under restrictive assumptions. For instance, in the three-dimensional atmospheric entry problem [9], the use of directly matched asymptotic expansions leads to a set of transcendental equations which can only be solved by resorting to numerical methods.

This Chapter addresses a three-dimensional atmospheric entry problem, to be analyzed by the method of matched asymptotic expansions (MAE). The solution is expressed in three parts; outer, inner, and common solutions. The outer solution is valid in the region where gravity is predominant. On

---

This Chapter is based on D. S. Naidu, "Three-dimensional atmospheric entry problem using method of matched asymptotic expansions," IEEE Trans. Aerospace and Electronic Systems, Vol. 25, pp. 660-667, Sept. 1989. @ 1989 IEEE. Permission given by IEEE is hereby acknowledged.

the other hand, the aerodynamically predominant region gives an inner solution. Since these two regions are bound to overlap, a matching process is required to identify the common solution. Thus, a composite solution, valid in the entire region, is constructed as the sum of the outer solution and inner solution from which we need to subtract the common solution. The matching principle, in other words, ties the constants of integration associated with the outer and inner solutions with given auxiliary conditions. Compared to the earlier work [9], the present method has the following features: (i) Analytical expressions have been obtained explicitly for the outer, inner and common solutions without facing a set of transcendental equations which can only be solved by numerical methods. (ii) The composite solution satisfies the given auxiliary conditions asymptotically. (iii) The common solution is generated as a polynomial in the stretched variable without actually solving for it from the inner limit of the outer solution or the outer limit of the inner solution. The present method is applicable to obtain explicit solutions for autonomous guidance and control strategies for aerospace vehicles.

2.2 Equations Of Motion

Consider a vehicle with constant point mass m, moving about a nonrotating spherical planet. The atmosphere surrounding the planet is assumed to be at rest, and the central gravitational field obeys the usual inverse square law. The equations of motion for three dimensional flight of the lifting vehicle are given by [Fig. 1] [7,9,10,12],

$$\frac{dr}{dt} = V\sin\gamma$$

$$\frac{d\theta}{dt} = \frac{V\cos\gamma\cos\psi}{r\cos\phi}$$

$$\frac{d\phi}{dt} = \frac{V\cos\gamma\sin\psi}{r}$$

$$\frac{dV}{dt} = -\frac{D}{m} - g\sin\gamma$$

$$V\frac{d\gamma}{dt} = \frac{L\cos\sigma}{m} - \left(g - V^2/r\right)\cos\gamma$$

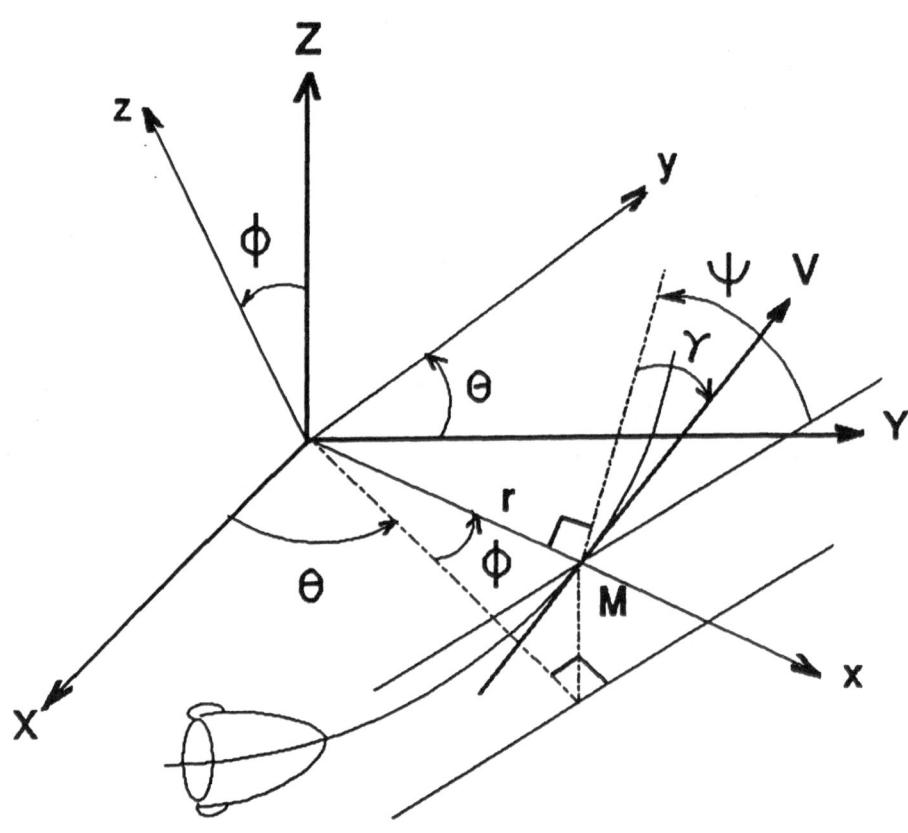

Fig. 1 Coordinate System

$$\frac{d\psi}{dt} = \frac{L\sin\sigma}{m\cos\gamma} - \frac{V^2}{r}\cos\gamma\cos\psi\tan\phi \qquad (1)$$

where it is assumed that the aerodynamic drag and lift are

$$D = \frac{1}{2}\rho SC_D V^2; \qquad L = \frac{1}{2}\rho SC_L V^2$$

and the gravitational field is

$$g = \frac{g_s r_s^2}{r^2} = \mu/r^2; \qquad \mu = g_s r_s^2$$

and the atmosphere is given by

$$\rho = \rho_s \exp\left[-\beta(r-r_s)\right]$$

For any particular flight program, the control functions C_L, C_D, and σ are given functions of time and the solution of (1) requires prescribing six initial conditions. It is convenient to eliminate time t in (1). Then, we get

$$\frac{d\theta}{dr} = \frac{\cos\psi}{r\cos\phi\tan\gamma}$$

$$\frac{d\phi}{dr} = \frac{\sin\psi}{r\tan\gamma}$$

$$\frac{dV^2}{dr} = -\frac{\rho SC_D V^2}{m\sin\gamma} - 2g$$

$$\frac{d\gamma}{dr} = \frac{\rho SC_L\cos\sigma}{2m\sin\gamma} - (g/V^2 - 1/r)\cot\gamma$$

$$\frac{d\psi}{dr} = \frac{\rho SC_L \sin\sigma}{2ms\sin\gamma\cos\gamma} - \frac{\cos\psi\tan\phi}{r\tan\gamma} \tag{2}$$

Solution of the set of five first order nonlinear differential equations (2) requires integration by numerical methods. Here the aim is to obtain approximate analytical solutions to (2) using some simplifications in the method of MAE.

2.3 Method Of Matched Asymptotic Expansions

In applying the method of MAE to the three-dimensional entry problem, consider separately the flight in an outer region near the vacuum, where the gravity force dominates, and an inner region near the planetary surface where the aerodynamic force is predominant. There is bound to be an overlap or common region where both outer and inner solutions are approximately of equal strength. A matching principle is invoked to obtain the common part. An approximate solution called the composite solution valid over the entire region, is constructed from the outer, inner, and common solutions. The various components of the composite solution are separated depending on the altitude, and hence it is appropriate to choose the altitude as an independent variable for obtaining the solution of (2).

Define the following dimensionless quantities,

$$h = (r-r_s)/r_s; \quad v = V^2/g_s r_s; \quad \varepsilon = 1/\beta r_s$$

$$B = \rho_s SC_D/2m\beta; \quad \lambda = C_L/C_D$$

Here the constant βr_s is large, i.e., for Earth's atmosphere $\beta r_s = 900$, and hence the parameter ε is a small quantity. Using the above dimensionless quantities in (2),

$$\frac{d\theta}{dh} = \frac{\cos\psi\cot\gamma}{(1+h)\cos\phi}$$

$$\frac{d\phi}{dh} = \frac{\sin\psi\cot\gamma}{(1+h)}$$

$$\frac{dv}{dh} = -\frac{2Bv\exp(-h/\varepsilon)}{\varepsilon\sin\gamma} - \frac{2}{(1+h)^2}$$

$$\frac{d\gamma}{dh} = \frac{B\lambda\cos\sigma\exp(-h/\varepsilon)}{\varepsilon\sin\gamma} + \left[\frac{1}{(1+h)} - \frac{1}{v(1+h)^2}\right]\cot\gamma$$

$$\frac{d\psi}{dh} = \frac{B\lambda\sin\sigma\exp(-h/\varepsilon)}{\varepsilon\sin\gamma\cos\gamma} - \frac{\cos\psi\tan\phi\cot\gamma}{(1+h)} \tag{3}$$

Although (3) is ready for analysis by the method of MAE, it is more convenient to replace the set of variables (θ, ϕ, ψ) with a new set of variables (α, Ω, I) which are related as [Fig. 2] [9,10],

$$\left.\begin{array}{l}\cos I = \cos\phi\cos\psi \\ \sin\phi = \sin I \sin\alpha \\ \cos\alpha = \cos\phi\cos(\theta-\Omega)\end{array}\right\} \tag{4}$$

where I is the inclination of the plane of the osculating orbit, Ω is the longitude of the ascending node, and α is the angle between the line of the ascending node and the position vector. Using (4) in (3), we get

$$\frac{d\alpha}{dh} = -\frac{\sin\alpha}{\tan I}\left\{\frac{B\lambda\sin\sigma\exp(-h/\varepsilon)}{\varepsilon\sin\gamma\cos\gamma}\right\} + \frac{\cot\gamma}{(1+h)}$$

$$\frac{d\Omega}{dh} = \frac{\sin\alpha}{\sin I}\left\{\frac{B\lambda\sin\sigma\exp(-h/\varepsilon)}{\varepsilon\sin\gamma\cos\gamma}\right\}$$

$$\frac{dI}{dh} = \cos\alpha\left\{\frac{B\lambda\sin\sigma\exp(-h/\varepsilon)}{\varepsilon\sin\gamma\cos\gamma}\right\}$$

$$\frac{dv}{dh} = -\frac{2Bv\exp(-h/\varepsilon)}{\varepsilon\sin\gamma} - \frac{2}{(1+h)^2}$$

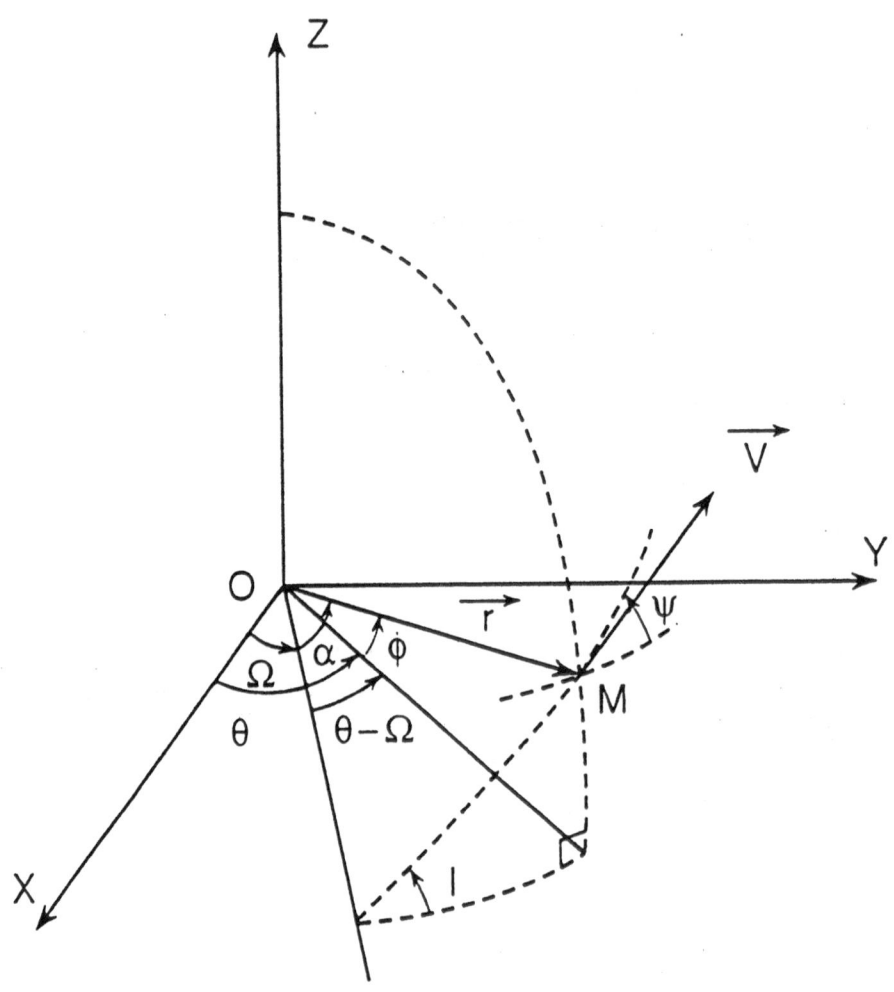

Fig. 2   Orbital Elements in Keplerian Motion

$$\frac{d\gamma}{dh} = \frac{B\lambda\cos\sigma\exp(-h/\varepsilon)}{\varepsilon\sin\gamma} + \left[\frac{1}{(1+h)} - \frac{1}{v(1+h)^2}\right]\cot\gamma \qquad (5)$$

with initial conditions, α_i, Ω_i, I_i, v_i, and γ_i.

Now introduce separate expansions for the outer (Keplerian) region and the inner region. The outer and inner regions are characterized by dominance of the gravitational and aerodynamic forces respectively.

2.3.1 Outer (Keplerian) Region

The outer expansions describe the solution in the region near vacuum where the gravitational force is predominant. These are assumed as

$$\alpha = \alpha_o(h) + \varepsilon\alpha_1(h) + \ldots\ldots$$

and similarly the other functions Ω, I, v, and, γ are expanded in powers of ε. By substituting the outer expansions into the original set of equations (5), and equating coefficients of ε^o on either side, the set of equations for zeroth-order approximation is

$$\frac{d\alpha_o}{dh} = \frac{\cot\gamma_o}{(1+h)}$$

$$\frac{d\Omega_o}{dh} = 0$$

$$\frac{dI_o}{dh} = 0$$

$$\frac{dv_o}{dh} = -\frac{2}{(1+h)^2}$$

$$\frac{d\gamma_o}{dh} = \left[\frac{1}{(1+h)} - \frac{1}{v_o(1+h)^2}\right]\cot\gamma_o \qquad (6)$$

Let us note that the zeroth-order equations (6) are alternatively obtained by letting the small parameter ε tend to zero in (5). The effect of making ε $(=1/\beta r_s)$ zero is that the atmospheric density ρ {$= \rho_s \exp(-h/\varepsilon)$} becomes zero and the resulting equations (6) describe the region near the vacuum. Hence, to a first order, the outer limit physically corresponds to a vanishing atmosphere. Solving (6),

$$v_o = 2\left[C_1 + \frac{1}{(1+h)}\right]$$

$$\cos\gamma_o = \frac{C_2}{(1+h)\sqrt{v_o}}$$

$$\cos(\alpha_o - C_3) = \frac{C_2^2/(1+h)-1}{\sqrt{1+2C_1 C_2^2}}$$

$$\Omega_o = C_4$$

$$I_o = C_5 \tag{7}$$

where C_i are the constants of integration to be determined. The first and higher order solutions are all equal to zero because at high altitude, in the limit, the atmospheric density is zero and the motion is Keplerian.

It is to be noted that the control variables, lift (C_L), drag (C_D), and bank angle (σ), are absent from the above zeroth-order outer solutions. That is, these controls have no effect in the outer region. The physical interpretation of this situation is that at higher altitude, the atmospheric density is too low to provide enough lift and drag forces for influencing the motion.

2.3.2 Inner (Aerodynamic) Region

The inner expansions are introduced to study the limiting condition of the solution near the planetary surface where the aerodynamic force is predominant. These are obtained by first using a stretching transformation

$$\bar{h} = h/\varepsilon$$

in (5) and then taking the limit $\varepsilon \to 0$. This corresponds to the region near $h = 0$, i.e., planetary surface. Thus the stretched system becomes

$$\frac{d\bar{\alpha}}{d\bar{h}} = -\frac{\sin\bar{\alpha}}{\tan\bar{I}}\left[\frac{B\lambda\sin\sigma\exp(-\bar{h})}{\sin\bar{\gamma}\cos\bar{\gamma}}\right] + \frac{\varepsilon\cot\bar{\gamma}}{(1+\varepsilon\bar{h})}$$

$$\frac{d\bar{\Omega}}{d\bar{h}} = \frac{\sin\bar{\alpha}}{\sin\bar{I}}\left\{\frac{B\lambda\sin\sigma\exp(-\bar{h})}{\sin\bar{\gamma}\cos\bar{\gamma}}\right\}$$

$$\frac{d\bar{I}}{d\bar{h}} = \cos\bar{\alpha}\left\{\frac{B\lambda\sin\sigma\exp(-\bar{h})}{\sin\bar{\gamma}\cos\bar{\gamma}}\right\}$$

$$\frac{d\bar{v}}{d\bar{h}} = -\frac{2B\bar{v}\exp(-\bar{h})}{\sin\bar{\gamma}} - \frac{2\varepsilon}{(1+\varepsilon\bar{h})^2}$$

$$\frac{d\bar{\gamma}}{d\bar{h}} = \frac{B\lambda\cos\sigma\exp(-\bar{h})}{\sin\bar{\gamma}} + \varepsilon\left[\frac{1}{(1+\varepsilon\bar{h})} - \frac{1}{\bar{v}(1+\varepsilon\bar{h})^2}\right]\cot\bar{\gamma} \qquad (8)$$

Note that λ and σ, assumed to be external control inputs do not undergo transformation. Let the inner solution be expressed as

$$\bar{\alpha} = \bar{\alpha}_0(\bar{h}) + \varepsilon\bar{\alpha}_1(\bar{h}) + \text{.......}$$

$$\bar{\Omega} = \bar{\Omega}_0(\bar{h}) + \varepsilon\bar{\Omega}_1(\bar{h}) + \text{.......}$$

$$\bar{I} = \bar{I}_0(\bar{h}) + \varepsilon\bar{I}_1(\bar{h}) + \text{.......}$$

$$\bar{v} = \bar{v}_0(\bar{h}) + \varepsilon\bar{v}_1(\bar{h}) + \ ......$$

$$\bar{\gamma} = \bar{\gamma}_0(\bar{h}) + \varepsilon\bar{\gamma}_1(\bar{h}) + \ ......$$

As before, substitution of the above into (8) and collection of coefficients of powers of ε^0 on either side gives the zeroth-order approximation as

$$\frac{d\bar{\alpha}_0}{d\bar{h}} = -\ \frac{\sin\bar{\alpha}_0}{\tan\bar{I}_0}\left\{\frac{B\lambda\sin\sigma\exp(-\bar{h})}{\sin\bar{\gamma}_0\cos\bar{\gamma}_0}\right\}$$

$$\frac{d\bar{\Omega}_0}{d\bar{h}} = \frac{\sin\bar{\alpha}_0}{\sin\bar{I}_0}\left\{\frac{B\lambda\sin\sigma\exp(-\bar{h})}{\sin\bar{\gamma}_0\cos\bar{\gamma}_0}\right\}$$

$$\frac{d\bar{I}_0}{d\bar{h}} = \cos\bar{\alpha}_0\left\{\frac{B\lambda\sin\sigma\exp(-\bar{h})}{\sin\bar{\gamma}_0\cos\bar{\gamma}_0}\right\}$$

$$\frac{d\bar{v}_0}{d\bar{h}} = -\ \frac{2B\bar{v}_0\exp(-\bar{h})}{\sin\bar{\gamma}_0}$$

$$\frac{d\bar{\gamma}_0}{d\bar{h}} = \frac{B\lambda\cos\sigma\exp(-\bar{h})}{\sin\bar{\gamma}_0}$$

Solving the above,

$$\bar{v}_0 = \bar{C}_1\exp\left(-2\bar{\gamma}_0/\lambda\cos\sigma\right)$$

$$\cos\bar{\gamma}_0 = B\lambda\cos\sigma\exp(-\bar{h}) + \bar{C}_2$$

$$\sin\bar{\alpha}_o \sin\bar{I}_o = \sin\bar{C}_3$$

$$\cos\bar{\alpha}_o = \cos\bar{C}_3\cos\left(\bar{C}_4 - \bar{\Omega}_o\right)$$

$$\cos\bar{I}_o = \cos\bar{C}_3\cos\left\{\tan\bar{\sigma}\log\left[\tan(\bar{\gamma}_o/2+\pi/4)\right]+\bar{C}_5\right\} \tag{9}$$

where $\bar{C}_i$ are the constants of integration. Note that the control variables, lift (C_L), drag (C_D), and bank angle (σ) are present in the zeroth-order inner solution (9). The physical meaning of this is that the zeroth-order inner solutions correspond to regions near the surface of the planet and these solutions are influenced by the aerodynamic controls.

Now there are the zeroth-order outer solutions (7) with C_i as the constants of integration and the zeroth-order inner solutions (9), where $\bar{C}_i$ are the integration constants. These constants are determined by a matching principle.

2.4 Matching Principle

Matching is based on the notion that the outer solution valid in the Keplerian region and the inner solution valid near the planet surface, must both be valid in some overlap region. Thus matching is accomplished by extending the outer solution into the inner region by transforming the outer variable h to that of the inner variable $\hat{h}$ (=h/ε) and taking the limit as ε → 0. This is called the inner limit of the outer solution or expansion. Similarly, the outer limit of the inner solution or expansion is obtained by extending the inner solution into the outer region by transforming the inner variable $\hat{h}$ to that of the outer variable h (= εh) and taking the limit as ε → 0. By equating the inner limit of outer expansion with the outer limit of inner expansion, we can determine the constants of integration and hence the common solution. A composite solution is formed as the sum of outer and inner solutions from which the common solution is subtracted.

In the earlier work [9], the matching principle yielded a relation for the constants $\bar{C}_i$ in terms of the constants C_i. Then the composite solution is expected to satisfy the given initial conditions. This procedure led to the formulation of a set of transcendental equations which can only be solved by resorting to numerical methods.

In the present method, the procedure is simplified by asking the outer solution to satisfy the given initial conditions and the matching principle gives the relation between the constants of integration [4]. Still, the composite solution satisfies the given initial conditions asymptotically. Note that in the simplified procedure, any kind of transcendental equations are avoided and explicit solutions are obtained for the composite solution. Moreover, the common solution is very easily obtained by formulating or generating the various terms of the inner limit of the outer expansion as a polynomial in the stretched variable as [14,15],

$$v_o^m = v_o(h=0) + \epsilon[v_1(h=0) + h\bar{v}_o(h=0)] + \ldots\ldots$$

$$\gamma_o^m = \gamma_o(h=0) + \epsilon[\gamma_1(h=0) + \bar{h}\dot{\gamma}_o(h=0)] + \ldots\ldots$$

$$\alpha_o^m = \alpha_o(h=0) + \epsilon[\alpha_1(h=0) + \bar{h}\dot{\alpha}_o(h=0)] + \ldots\ldots$$

$$\Omega_o^m = \Omega_o(h=0) + \epsilon[\Omega_1(h=0) + \bar{h}\dot{\Omega}_o(h=0)] + \ldots\ldots$$

$$I_o^m = I_o(h=0) + \epsilon[I_1(h=0) + \bar{h}\dot{I}_o(h=0)] + \ldots\ldots\ldots$$

Here the dot denotes differentiation with respect to the independent variable h. Note that this is also called the intermediate solution in singular perturbation methods [14, 15].

Now force the outer solution (7) to satisfy the given initial conditions, v_i, γ_i, α_i, Ω_i, and I_i corresponding to $h = h_i$. This gives

$$C_1 = v_i/2 - 1/(1+h_i)$$

$$C_2 = \cos\gamma_i(1+h_i)\sqrt{v_i}$$

$$C_3 = \alpha_i - \cos^{-1}\left\{\frac{\cos^2\gamma_i(1+h_i)v_i - 1}{\sqrt{1+[v_i(1+h_i)-2]}[\cos^2\gamma_i(1+h_i)v_i]}\right\}$$

$$C_4 = \Omega_i$$

$$C_5 = I_i \tag{10}$$

Thus, the relation between the constants of outer solution explicitly in terms of the given initial conditions is obtained. In applying the matching principle, first find the zeroth-order inner limit of the outer expansion (7), as

$$v_0^m = 2(C_1+1)$$

$$\cos\gamma_0^m = C_2/\sqrt{2(C_1+1)}$$

$$\cos(\alpha_0^m-C_3) = (C_2^2-1)/\sqrt{1+2C_1C_2^2}$$

$$\Omega_0^m = C_4$$

$$I_0^m = C_5 \tag{11}$$

The outer limit of inner expansion (9) is

$$\bar{v}_0^{-m} = \bar{C}_1\exp\left(-2\bar{\gamma}_0^{-m}/\lambda\cos\sigma\right)$$

$$\cos\bar{\gamma}_0^{-m} = \bar{C}_2$$

$$\sin\bar{\alpha}_o^{-m}\sin\bar{I}_o^{-m} = \sin\bar{C}_3$$

$$\cos\bar{\alpha}_o^{-m} = \cos\bar{C}_3\cos(\bar{C}_4-\bar{\Omega}_o^{-m})$$

$$\cos\bar{I}_o^{-m} = \cos\bar{C}_3\cos\left\{\tan\sigma\log\left[\tan(\bar{\gamma}_o^{-m}/2+\pi/4)\right]+\bar{C}_5\right\} \tag{12}$$

Matching (12) with (11) gives constants $\bar{C}_i$ in terms of the constants C_i as

$$\bar{C}_1 = 2(C_1+1)\exp\left\{\frac{2}{\lambda\cos\sigma}\cos^{-1}\left[C_2/\sqrt{2(C_1+1)}\right]\right\}$$

$$\bar{C}_2 = C_2/\sqrt{2(C_1+1)}$$

$$\sin\bar{C}_3 = \sin\left[C_3+\cos^{-1}\left\{(C_2^2-1)/\sqrt{1+2C_1C_2^2}\right\}\right]\sin C_5$$

$$\bar{C}_4 = C_4 + \cos^{-1}\left[\cos\left[C_3+\cos^{-1}\left\{(C_2^2-1)/\sqrt{1+2C_1C_2^2}\right\}\right]/\cos\bar{C}_3\right]$$

$$\bar{C}_5 = \cos^{-1}\left[\cos C_5/\cos\bar{C}_3\right] - \tan\sigma\log\left[\tan\left\{\pi/4+\cos^{-1}\left[C_2/\sqrt{8(C_1+1)}\right]\right\}\right]$$

Note that the above constants $\bar{C}_i$ are in turn related with the constants C_i.

2.5 Composite Solution

In the previous sections, analytical solutions are obtained if flight conditions are considered separately in two widely different regions; outer region at high altitude where aerodynamic force is weak compared with gravitational force and the inner region near the planetary surface where aerodynamic force is predominant over the gravitational force.

Now construct a composite solution which is valid everywhere in the gravitationally dominant outer region and aerodynamically strong inner region. The composite solution or expansion is obtained as the sum of outer solution (7) and inner solution (9) from which the common solution (11) or (12) is subtracted. Thus

$$v_c = v_0 + \bar{v}_0 - v_0^m(\text{ or } \bar{v}_0^m)$$

$$\gamma_c = \gamma_0 + \bar{\gamma}_0 - \gamma_0^m(\text{ or } \bar{\gamma}_0^m)$$

$$\alpha_c = \alpha_0 + \bar{\alpha}_0 - \alpha_0^m(\text{ or } \bar{\alpha}_0^m)$$

$$\Omega_c = \Omega_0 + \bar{\Omega}_0 - \Omega_0^m(\text{ or } \bar{\Omega}_0^m)$$

$$I_c = I_0 + \bar{I}_0 - I_0^m(\text{ or } \bar{I}_0^m) \tag{13}$$

In terms of explicit expansions,

$$v_c = 2/(1+h) + \check{C}_1 \exp\left(-2\bar{\gamma}_0/\lambda\cos\sigma\right) - 2.0$$

$$\gamma_c = \cos^{-1}\left\{\frac{C_2}{(1+h)\sqrt{2C_1+2/(1+h)}}\right\} + \cos^{-1}\left\{B\lambda\cos\sigma\exp(-h/\epsilon)+\check{C}_2\right\}$$

$$- \cos^{-1}\left\{C_2/\sqrt{2(C_1+1)}\right\}$$

$$\Omega_c = \check{C}_4 - \cos^{-1}\left[\cos\bar{\alpha}_0/\cos\check{C}_3\right]$$

$$\cos I_c = \cos\check{C}_3\cos\left[\tan\sigma\log\left(\tan(\bar{\gamma}_0/2+\pi/4)\right)+\check{C}_5\right]$$

$$\alpha_c = \cos^{-1}\left\{\frac{C_2^2/(1+h)-1}{\sqrt{1+2C_1C_2^2}}\right\} + \sin^{-1}\left\{\frac{\sin\bar{C}_3}{\sin\bar{I}_o}\right\} - \cos^{-1}\left\{\frac{C_2^2-1}{\sqrt{1+2C_1C_2^2}}\right\} \tag{14}$$

The above composite solution is expressed in terms of h, constants $\bar{C}_i$, and C_i and the states from the inner solution. The states of inner solution are obtained as explicit functions of the $\bar{C}_i$ and $\bar{h}$ via (9).

To check whether the composite solution (14) asymptotically satisfies the given initial conditions, consider (14) along with (9), (11). Then

$$v_c(h=h_i) = \frac{2}{(1+h_i)} + 2(C_1+1)\exp\left\{\frac{2}{\lambda\cos\sigma}\cos^{-1}\left[C_2/\sqrt{2(C_1+1)}\right]\right\}$$

$$\exp\left[-\frac{2}{\lambda\cos\sigma}\cos^{-1}\left\{B\cos\sigma\exp(-h_i/\varepsilon)+C_2/\sqrt{2(C_1+1)}\right\}\right] - 2.0$$

As $\varepsilon \to 0$

$$v_c(h=h_i) = \frac{2}{(1+h_i)} + 2C_1 = v_i$$

Similarly, it can be shown that γ_c, I_c, α_c, and Ω_c satisfy their corresponding initial conditions asymptotically.

2.6 Concluding Remarks

In this Chapter, the solution of a three-dimensional atmospheric entry problem via a simplified method of matched asymptotic expansions has been addressed. The solution has been expressed in three parts. An outer solution has been obtained in the gravitationally dominant region and an inner solution has been formed for the aerodynamically stronger region. A common solution has been formed as the outer (or inner) limit of the inner (or outer) solution. Finally, a composite solution has been constructed as the sum of the outer and inner solutions from which the common solution has been subtracted.

The special features of the present method are (i) The composite solution has been obtained in a simplified manner in the sense that analytical expressions have been obtained explicitly for the various components, outer, inner and common solutions, without resorting to any kind of transcendental equations which can only be solved by numerical methods. (ii) At the same time, the composite solution has satisfied the given initial conditions asymptotically. (iii) The common solution has been obtained in an easier manner by formulating or generating the inner limit of the outer solution in terms of a polynomial in the stretched variable.

With respect to the applications spectrum, first note that the method of matched asymptotic expansions is a powerful technique in obtaining approximate, explicit solutions for nonlinear dynamics describing a variety of flight conditions for aerospace vehicles and that on board strategies are critical to any mission success. These simplified, analytical solutions are useful for on board guidance and control because of the fact that the off board (ground-based) computations are unable to confront problems such as imperfect modeling of vehicles, and atmospheric uncertainties during actual flight. Also, in optimization (minimum fuel or maximum payload) problems, one is faced with a tremendous computational task of solving on board nonlinear two-point boundary value problems. The present methodology which offers simplified computations is immensely suitable for such on board situations.

Nomenclature

C_i : constants of integration for outer solution

$\bar{C}_i$: constants of integration for inner solution

C_D :   drag coefficient

C_L : lift coefficient

D :   drag force

g : gravitational acceleration

g_s : gravitational acceleration at surface level

h : nondimensional altitude

I : inclination of the orbital plane

L : lift force

m :   vehicle mass

r : distance of the vehicle from planet center

r_s : radius of the planet at surface level

S : aerodynamic reference area

t : time

V : velocity

v : nondimensional velocity
α : angle between the ascending node and the position vector
β : inverse atmospheric scale height
γ : flight path angle
ψ : heading angle
σ : bank angle
θ : down range angle or longitude
φ : cross range angle or latitude
ρ : density
μ : gravitational constant of Earth
Ω : longitude of the ascending node

References

[1] Y. C. Shen, "Series solution of equations for re-entry vehicles with variable lift and drag coefficients," *AIAA Journal*,1, 2487-2490, 1963.

[2] Y. Y. Shi and M. C. Eckstein, "Ascent or descent from satellite orbit by low thrust," *AIAA Journal*, 4, 2203-2209, 1966.

[3] R. E. Willes, M. C. Francisco, J. G. Reid, and W. K. Lim, "An application of matched asymptotic expansions to hypervelocity flight mechanics," AIAA Guidance, Control, and Flight Mechanics Conference, Huntsville, AL, Aug. 1967.

[4] Y.-Y. Shi and L. Pottsepp, "Asymptotic expansions of a hypervelocity atmospheric entry problem," *AIAA Journal*, 7, 353-355, 1969.

[5] Y.-Y. Shi, L. Pottsepp, and M. C. Eckstein, "A matched asymptotic solution for skipping entry into planetary atmosphere," *AIAA Journal*, 9, 736-738, 1971.

[6] Y.-Y. Shi, "Matched asymptotic solutions for optimum lift controlled atmospheric entry," *AIAA Journal*, 9, 2229-2238, 1971.

[7] E. D. Dickmanns, "Maximum range three-dimensional lifting planetary entry," NASA Tech. Report, TR R-387, Marshall Space Flight Center, AL, Aug. 1972

[8] R. V. Ramnath, and P. Sinha, "Dynamics of the space shuttle during entry into Earth's atmosphere," *AIAA Journal*, 13, 337-342, 1975.

[9] A. Busemann, N. X. Vinh, and R. D. Culp, "Solution of the exact equations for three-dimensional atmospheric entry using directly matched

asymptotic expansions," NASA Contractor Report, CR-2643, University of Colorado, Boulder, CO, March 1976.

[10] F. Frostic and N. X. Vinh, "Optimal aerodynamic control by matched asymptotic expansions," *Acta Astronautica*, 3, 319-332, 1976.

[11] K. D. Mease, and F. A. McGreary, "Atmospheric guidance law for planar skip trajectories," AIAA 12th Atmospheric Flight Mechanics Conference, Snowmass, CO, Aug. 1985.

[12] N. X. Vinh, *Optimal Trajectories in Atmospheric Flight*, Elsevier Scientific Publishing Co., Amsterdam, 1981.

[13] J. D. Cole, *Perturbation Methods in Applied Mathematics*, Blaisdell Publishing Co., Waltham, 1968.

[14] D. S. Naidu and D. B. Price, "On the method of matched asymptotic expansions," Journal of Guidance, Control, and Dynamics, 12, 277-279, 1989.

[15] D. S. Naidu, *Singular Perturbation Methodology in Control Systems*, IEE Control Engineering Series, Peter Peregrinus Limited, Stevenage Herts, England, 1988.

****

CHAPTER 3

FUEL OPTIMAL CONTROL FOR
COPLANAR ORBITAL TRANSFER

3.1 Introduction

In space transportation systems, the idea of aeroassisted orbital transfer opens new mission opportunities, especially with regard to the initiation of a permanent space station [1]. The use of aeroassisted maneuvers to affect a transfer from high Earth orbit (HEO) to low Earth orbit (LEO) has been recommended to provide high performance leverage to future space transportation systems. The space-based orbit transfer vehicle (OTV) is planned as a system for transporting payloads between LEO and other locations in space. For example, an OTV which is initially at a low Earth orbit, say the space station orbit (SSO), is required to transfer a payload to a high Earth orbit such as geosynchronous orbit, pick up another payload, say a faulty satellite and return to orbiting hanger located at SSO for refurbishment and redeployment. The OTV, on its return journey from HEO to LEO needs to dissipate its orbital energy . This can be accomplished by using entirely propulsive (Hohmann) transfer in space only or a combination of propulsive transfer in space and aeroassisted maneuver in atmosphere. It has been established that significant fuel savings of about 50% and hence increased payload capabilities can be achieved by using aeroassisted maneuver instead of the all-propulsive Hohmann transfer [2]. The aeroassisted orbital transfer vehicle (AOTV) utilizes atmospheric drag to slow down to LEO velocity instead of using rocket fuel for retrobraking. Thus, optimization of fuel is an important aspect in orbital transfer missions. Basically, it is a synergetic maneuver for an AOTV, employing a hybrid combination of propulsive maneuvers in space and aerodynamic maneuvers in sensible atmosphere [2-4].

This Chapter first describes briefly the various types of coplanar

---

This Chapter is based on D. S. Naidu, et. al., "Fuel-optimal trajectories for aeroassisted coplanar orbital transfer problem," IEEE Trans. Aerospace and Electronic Systems, Vol. 26, pp. 374-381, March 1980. @ 1990 IEEE. Permission given by IEEE is hereby acknowledged.

transfers [5]. Then addresses the fuel-optimal control problem arising in coplanar orbital transfer employing aeroassist technology. The maneuver involves a transfer from HEO to LEO and at the same time minimization of the fuel consumption for achieving the desired orbit transfer. It is known that a change in velocity, also called the characteristic velocity, is a convenient parameter to measure the fuel consumption. A suitable performance index is the total characteristic velocity which is the sum of the characteristic velocities for deorbit and for reorbit (or circularization) [6]. Use of Pontryagin minimum principle leads to a nonlinear, two-point boundary value problem (TPBVP) in state and costate variables. The solution of the TPBVP is a stumbling block in obtaining fuel-optimal solution. This problem is solved by using a more efficient multiple shooting method which is a simultaneous application of a single shooting algorithm to equally divided points of the total interval of the solution [7,8]. The fuel-optimal solution forms as a basis for on board guidance of an AOTV.

3.2 Types Of Coplanar Orbital Transfers

In transferring a vehicle from one orbit, say HEO to another orbit, say LEO, in the same plane, there are three types of orbital transfer: (i) Hohmann transfer, (ii) Idealized aeroassisted transfer, and (iii) Realistic aeroassisted transfer.

3.2.1 Hohmann Transfer

This is a coplanar transfer in which the vehicle is transferred from HEO to LEO in an all-propulsive manner [Fig. 1]. The first tangential impulse ΔV_d, called the deorbit impulse, at the radius R_d of HEO, is executed to transfer the vehicle from circular orbit to elliptic transfer orbit with its perigee at R_c. The second tangential impulse ΔV_c, called the circularization or reorbit impulse, at the radius R_c of LEO, is imparted to transfer the elliptic orbit to circular orbit. Thus, the total characteristic velocity ΔV_h, a measure of fuel consumption in achieving the Hohmann transfer, is given by the sum of the deorbit impulse ΔV_d, and the circularization impulse ΔV_c. Using the principle of conservation of angular momentum, the expressions for the deorbit and circularization impulses are given by [5]

$$\Delta V_{dh} = \sqrt{\mu/R_d}\left[1 - \sqrt{2/(1+R_d/R_c)}\right]$$

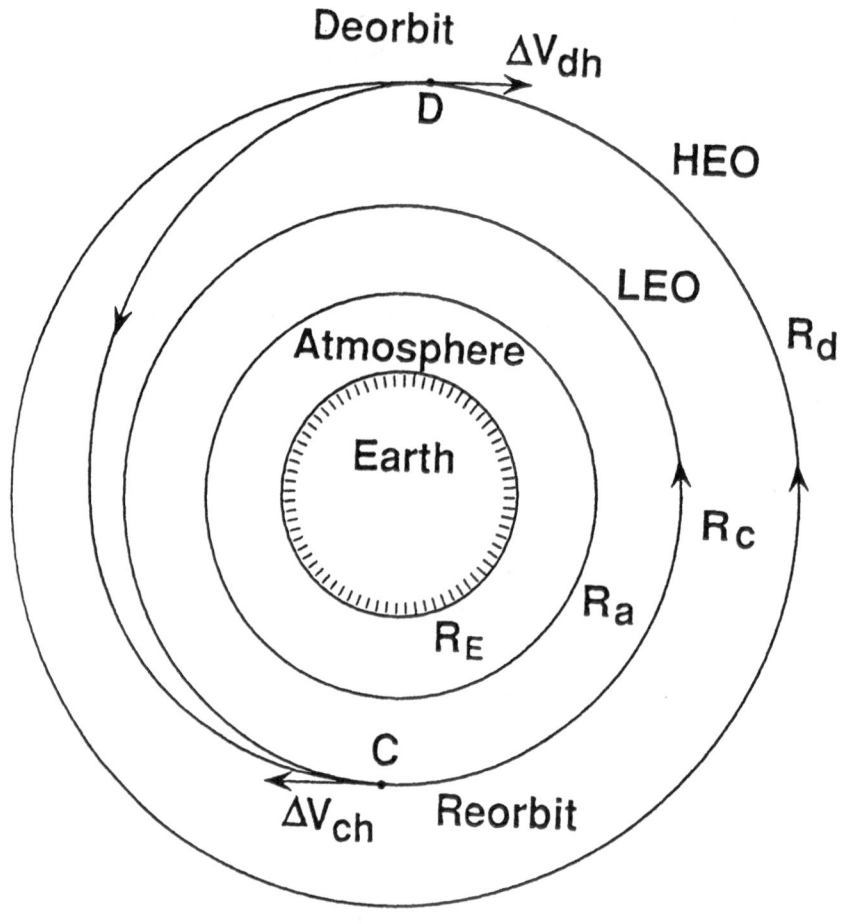

Fig. 1   Hohmann Transfer

$$\Delta V_{ch} = \sqrt{\mu/R_c}\left[\sqrt{2(R_d/R_c)/(1+R_d/R_c)} - 1\right] \tag{1}$$

Using the normalized values,

$$a_d = R_d/R_a; \quad a_c = R_c/R_a; \quad \Delta v = \Delta V/\sqrt{\mu/R_a}$$

in (1) gives

$$\Delta v_{dh} = \sqrt{1/a_d} - \sqrt{2a_c/a_d(a_d+a_c)}$$

$$\Delta v_{ch} = \sqrt{2a_d/a_c(a_d+a_c)} - \sqrt{1/a_c} \tag{2}$$

so that the total characteristic velocity in the normalized form is given by

$$\Delta v_h = \Delta v_{dh} + \Delta v_{ch}$$

3.2.2 Ideal Aeroassisted Transfer

In an idealized aeroassisted coplanar transfer, the vehicle grazes the atmospheric boundary, undergoes the necessary velocity reduction and skips back into another orbit [Fig. 2]. Thus, the vehicle leaves HEO at R_d with a tangential deorbit impulse ΔV_{di} and enters into an elliptic orbit with its perigee at R_a and flight path angle $\gamma_e = 0$. When the vehicle is at perigee E, its lifting capability (in this case negative lift) is employed to affect flight path along the boundary of the atmosphere (i.e., along the circular orbit of radius R_a). This grazing flight is continued along the atmospheric boundary to point F until sufficient velocity has been depleted by atmospheric drag such that upon reducing the lift to zero, the vehicle ascends with $\gamma_f = 0$ on an elliptic orbit to an apogee at R_c. Finally, at point C, a tangential circularizing burn ΔV_{ci} is imparted to achieve the desired LEO. In this transfer, the idealizations are (i) the atmospheric density at R_a is sufficient to generate enough drag to slow the vehicle in a reasonable amount of time, and (ii) the vehicle has sufficient lifting capability to maintain flight along the atmospheric boundary.

The following relations for the characteristic velocities are obtained by using the principles of conservation of angular momentum and the conservation of energy. The deorbit impulse ΔV_{di} is given by

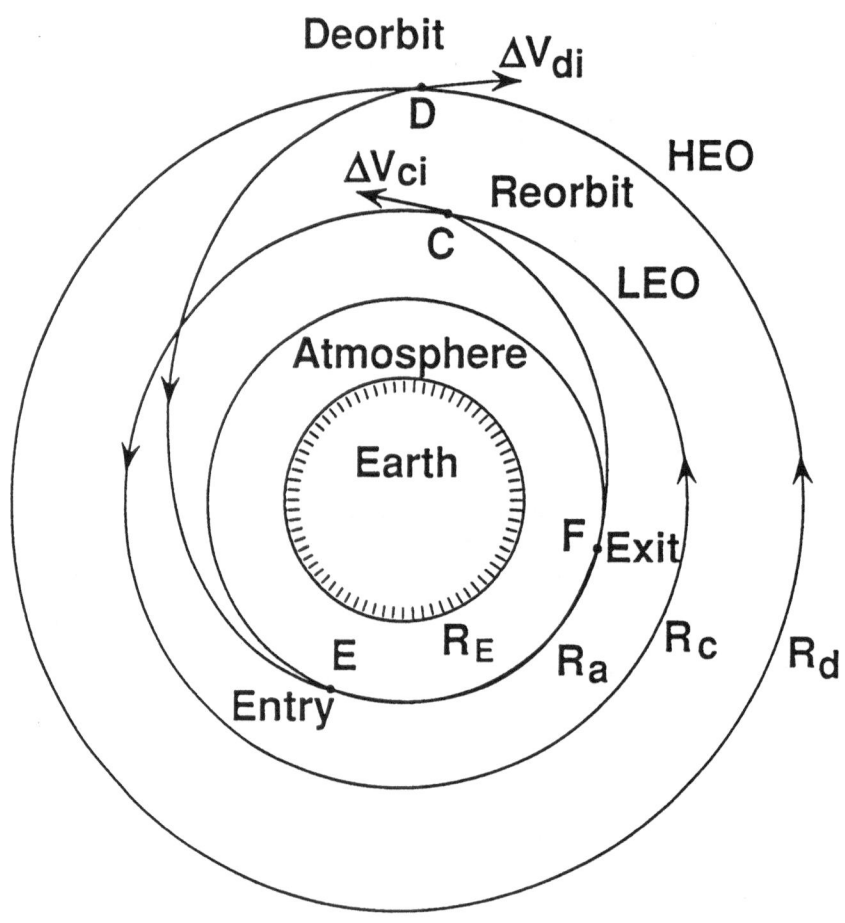

Fig. 2   Ideal Aeroassisted Transfer

$$\Delta V_{di} = \sqrt{\mu/R_d} - \sqrt{2(\mu/R_a)/(R_d/R_a)(R_d/R_a+1)}$$

and the circularization or reorbit impulse ΔV_{ci} turns out to be

$$\Delta V_{ci} = \sqrt{\mu/R_c} - \sqrt{2(\mu/R_a)/(R_c/R_a)(R_c/R_a+1)}$$

in terms of the normalized values

$$\Delta v_{di} = \sqrt{1/a_d} - \sqrt{2/a_d(a_d+1)}$$

$$\Delta v_{ci} = \sqrt{1/a_c} - \sqrt{2/a_c(a_c+1)} \tag{3}$$

and the total characteristic velocity for the idealized transfer in the normalized form is

$$\Delta v_i = \Delta v_{di} + \Delta v_{ci}$$

3.2.3 Realistic Aeroassisted Transfer

In a realistic, aeroassisted, coplanar transfer, the vehicle is transferred from HEO at R_d to LEO at R_c, by flying deep into the atmosphere to achieve the necessary velocity reduction [Fig. 3]. The transfer starts with a tangential propulsive burn, having a characteristic velocity ΔV_d for deorbiting from the high Earth orbit (HEO) and entering into an elliptical transfer orbit. At point E the spacecraft enters the atmosphere with an inclination of γ_e and undergoes reduction in velocity due to atmospheric drag. At point F, the spacecraft leaves the atmosphere with flight path angle γ_f. Once again, the transfer orbit is elliptical with the corresponding apogee at R_c. Finally, the maneuver ends with a circularizing or reorbit burn having a characteristic velocity ΔV_c to make the vehicle enter into the low Earth orbit. The desired circularization into LEO is achieved by the appropriate magnitude of ΔV_c. Thus, the maneuver consists of two impulses ΔV_d for deorbit, and ΔV_c for circularization and is assumed to take place right at the perigee itself. Upon applying the principles of conservation of angular momentum and the conservation of energy, the relations for the deorbit and the circularization impulses are obtained as

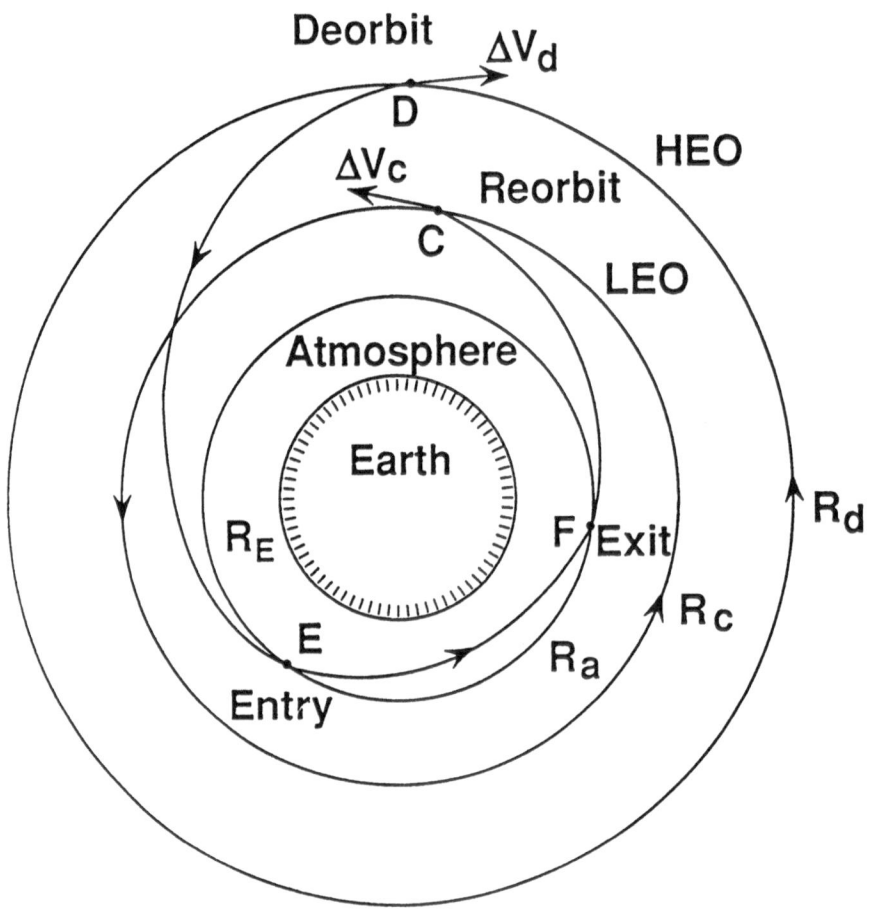

Fig. 3 Aeroassisted Coplanar Orbital Transfer

$$\Delta V_d = \sqrt{\mu/R_d}\left[1 - \sqrt{2(1-R_d/R_a)/\left(1-(R_d/R_a)^2/\cos^2\gamma_c\right)}\right]$$

$$\Delta V_c = \sqrt{\mu/R_c}\left[1 - \sqrt{2(1-R_c/R_a)/\left(1-(R_c/R_a)^2/\cos^2\gamma_f\right)}\right]$$

In terms of the normalized values, the above relations become

$$\Delta v_d = \sqrt{1/a_d} - \sqrt{2(1-a_d)/a_d(1-a_d^2/\cos^2\gamma_e)}$$

$$\Delta v_c = \sqrt{1/a_c} - \sqrt{2(1-a_c)/a_c(1-a_c^2/\cos^2\gamma_f)} \qquad (4)$$

Finally, the total characteristic velocity for the realistic aeroassisted transfer in the normalized form becomes

$$\Delta v = \Delta v_d + \Delta v_c$$

Note that if $\gamma_e = \gamma_f = 0$, the relations (4) for the various characteristic velocities for the realistic transfer are the same as those relations (3) for the corresponding characteristic velocities for the idealized transfer. Thus, the total characteristic velocity Δv for the realistic transfer is at least equal to that of Δv_i. In other words, the idealized transfer is the lower bound of the realistic transfer.

The entry velocity v_e, and the exit velocity v_f are also obtained in the normalized form as

$$v_e = \sqrt{2a_d(1-a_d)/(\cos^2\gamma_e - a_d^2)}$$

$$v_f = \sqrt{2a_c(1-a_c)/(\cos^2\gamma_f - a_c^2)} \qquad (5)$$

3.2.4 Comparison of Orbital Transfers

Now try to compare the three types of transfers, i.e., the Hohmann transfer, the ideal aeroassisted transfer, and the realistic aeroassisted transfer. For this, the following data are used. Radius of Earth, R_E = 6,378 km ;radius of atmosphere, R_a = 6,498 km; radius of the HEO, R_d = 42,241 km; radius of the LEO, R = 6,728 km; gravitational constant of Earth, μ = 3.986×10^{14} meter^3/sec^2. Using these data, the total characteristic velocities for each of the types of transfers are shown in Figure 4. See a clear advantage of using the aeroassisted transfer which has over 50 percent saving in fuel compared to the all propulsive Hohmann transfer. The variation of entry velocity V_e and the deorbit characteristic velocity ΔV_d with respect to entry flight path angle γ_e are depicted in Figure 5. From a given high Earth orbit, as the vehicle tries to dip deeper into atmosphere (i.e., to have lower perigee altitudes), it interfaces with atmosphere with a more negative flight path angle γ_e and less entry velocity V_e. Also, the apogee velocity decreases and hence the deorbit impulse ΔV_e increases. In other words, there is a need to impart higher deorbit impulse to enter deeper into atmosphere. The variation of exit velocity V_f and the reorbit characteristic velocity ΔV_f are shown in Figure 6. For reaching the same low Earth orbit, a higher positive exit flight path angle γ_f accompanied by a lower exit velocity V_f results in a lower apogee velocity at the circularization point. This leads to a higher reorbit impulse ΔV_c. The scenario at the entry and exit corridors shown in Figures 5 and 6 indicate that there is a need to fly the vehicle at near zero exit flight path angle for minimum fuel consumption.

3.3 Dynamics Of Motion

In an orbital transfer problem, the following assumptions are made for a vehicle with a constant point mass. (i) The initial HEO and final LEO are circular. (ii) It is a two-impulsive transfer with deorbiting impulse at HEO and reorbiting impulse at LEO. (iii) Within the atmosphere, the motion is planar,i.e., no plane change is assumed. (iv) The control is achieved only by lift modulation. (v) Earth's rotation is neglected. (vi) A Newtanion inverse square gravitational law is used. (vii) The vehicle has parabolic drag polar. Then, the equations of motion are given by,

$$\frac{dH}{dt} = V \sin\gamma$$

$$\frac{dV}{dt} = -AC_D V^2 \exp(-H\beta) - (\mu/R^2)\sin\gamma$$

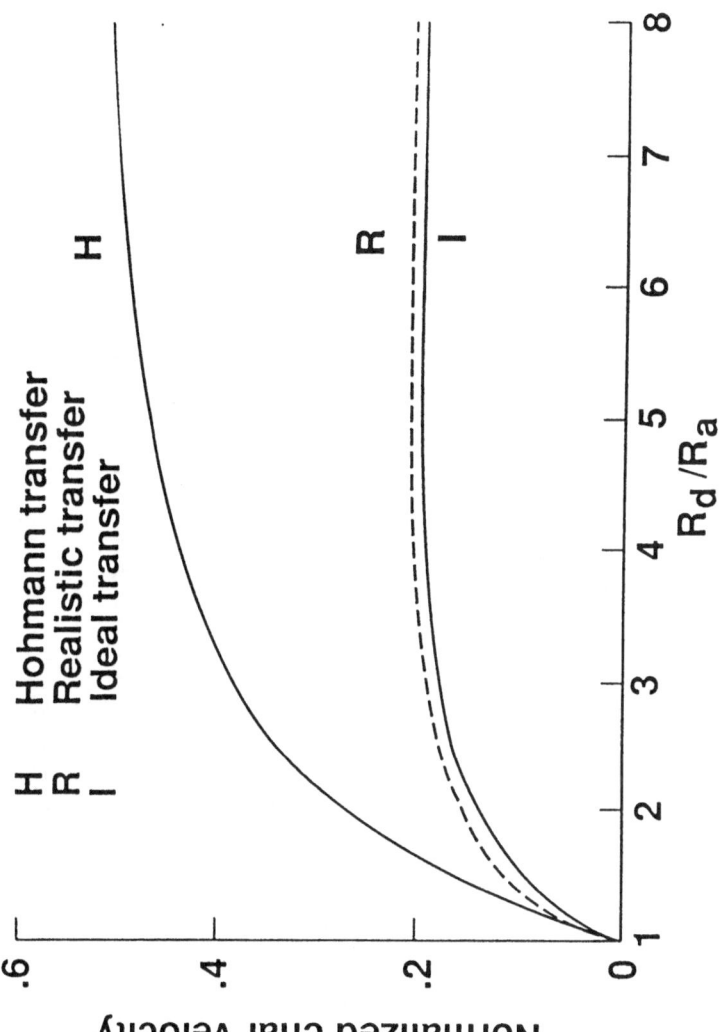

Fig. 4 Comparison of Orbital Transfers

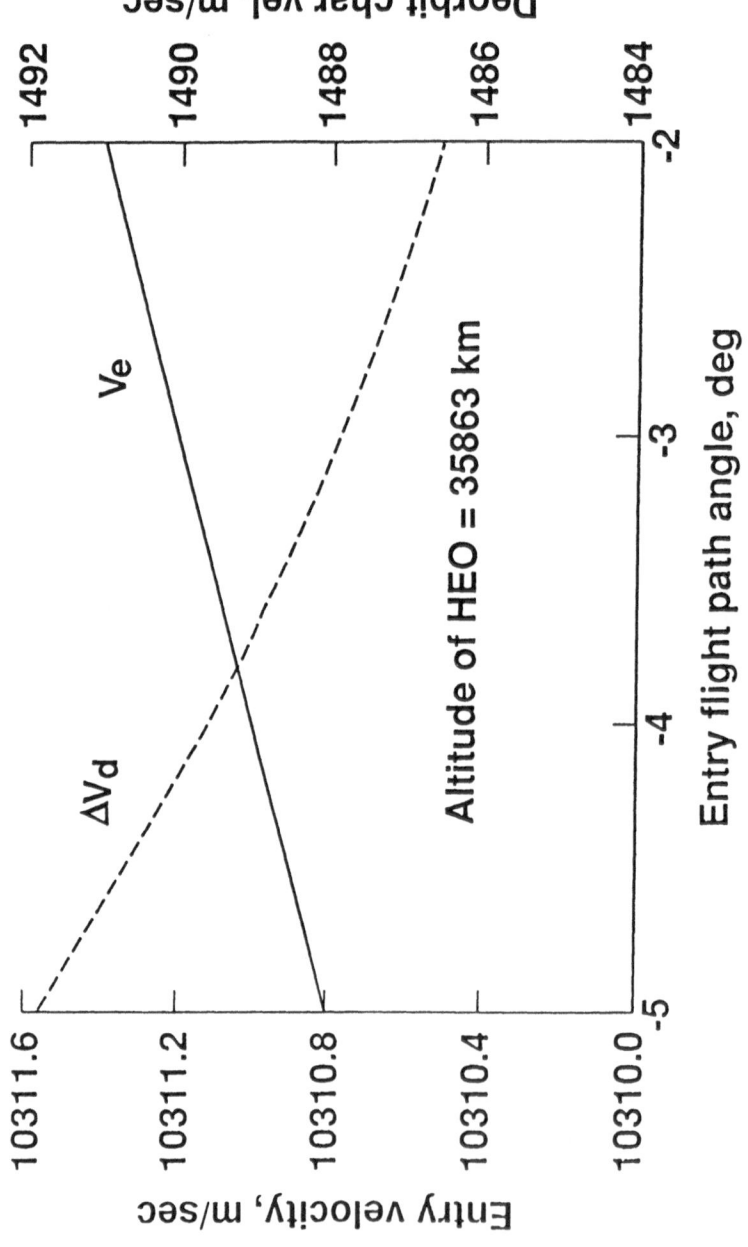

Fig. 5   Situation at Entry Corridor

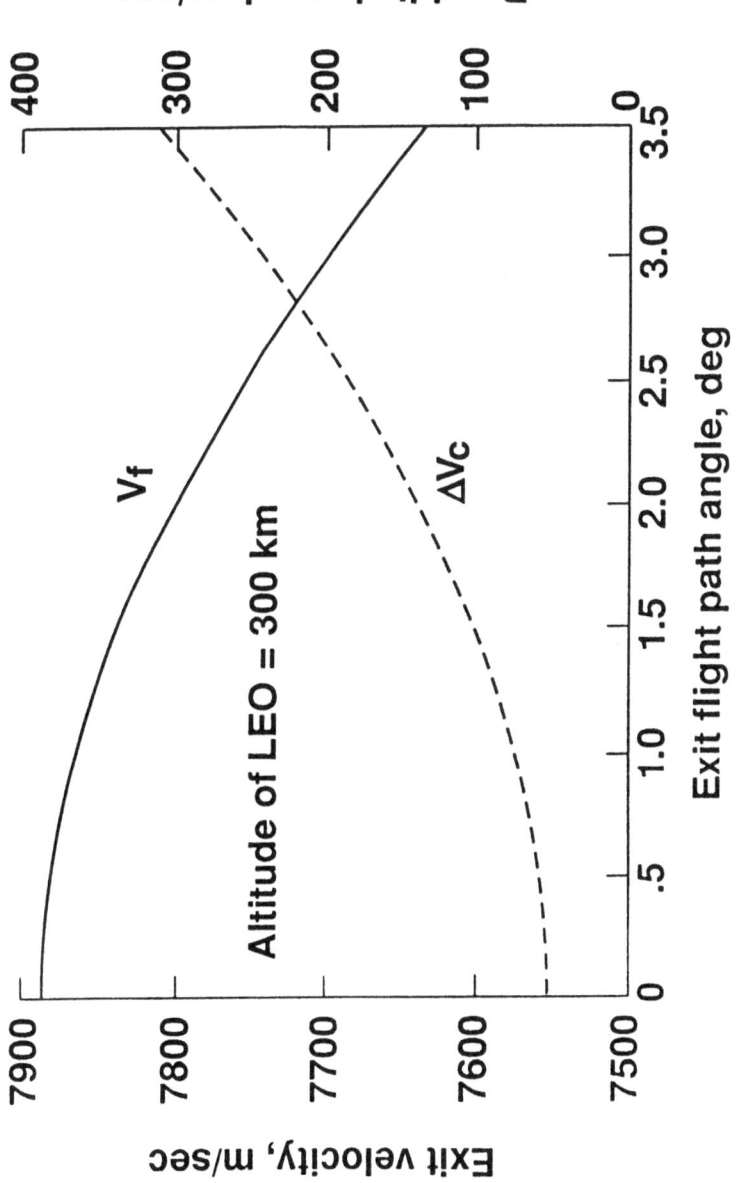

Fig. 6 Situation at Exit Corridor

$$\frac{d\gamma}{dt} = AC_L V \exp(-H\beta) + [V/R - \mu/(R^2 V)]\cos\gamma \tag{6}$$

where $A = S\rho/2m$, $H = R\text{-}R$, $\rho = \rho \exp(-H\beta)$ and

$$C_D = C_{D0} + KC_L^2.$$

Using the normalized variables,

$$\tau = t/\sqrt{R_a^3/\mu}; \quad v = V/\sqrt{\mu/R_a}$$

and the dimensionless constants,

$$h = H/H_a; \quad b = R_a/H_a; \quad \delta = \rho/\rho_s = \exp(-h\beta H_a)$$

$$\eta = C_L/C_{LR}; \quad C_{LR} = \sqrt{C_{D0}/K}$$

in (6), we get

$$\frac{dh}{d\tau} = bv\sin\gamma$$

$$\frac{dv}{d\tau} = -A_1 b(1+\eta^2)\delta v^2 - \frac{b^2\sin\gamma}{(b-1+h)^2}$$

$$\frac{d\gamma}{d\tau} = A_2 b\eta\delta v + \frac{bv\cos\gamma}{(b-1+h)} - \frac{b^2\cos\gamma}{(b-1+h)^2 v} \tag{7}$$

where, $A_1 = C_{D0}S\rho_s H_a/2m$; $A_2 = C_{LR}S\rho_s H_a/2m$

3.4 Optimal Control

For an optimal control problem with minimum fuel consumption, it is required to choose the performance index to minimize the total characteristic velocity, which is the sum of the initial characteristic velocity ΔV_d, the deorbit impulse from HEO, and the final characteristic velocity ΔV_c, the circularization impulse into LEO. Thus, the performance index is given by [6]

$$J = \Delta V = \Delta V_d + \Delta V_c$$

where,

$$\Delta V_d = \sqrt{\mu / R_d} - (R_a / R_d) V_e \cos(-\gamma_e)$$

$$\Delta V_c = \sqrt{\mu / R_c} - (R_a / R_c) V_f \cos \gamma_f$$

In the normalized form, the performance index becomes,

$$J = \Delta v = \Delta v_d + \Delta v_c$$

where,

$$\Delta v_d = \sqrt{1/a_d} - (v_e / a_d) \cos(-\gamma_e)$$

$$\Delta v_c = \sqrt{1/a_c} - (v_f / a_c) \cos(\gamma_f)$$

The minimization of J with respect to γ_e and γ_f yields $\gamma_e = 0$ and $\gamma_f = 0$. Because of this, it is natural to postulate that the optimal trajectory behaves as $\gamma(t) = 0$ for all values of time t between the entry time t_e and the exit time t_f. This constant flight path angle along the trajectory means that the altitude is constant at the value equal to altitude of the atmospheric interface. Thus, the optimal trajectory becomes an ideal (grazing) trajectory.

The interest is in finding the minimization of the fuel with respect to the control C_L. Using Pontryagin's principle, formulate the Hamiltonian as

$$\mathcal{H} = \lambda_h bv \sin\gamma + \lambda_v \left\{ -A_1 b(1+\eta^2)\delta v^2 - \frac{b^2 \sin\gamma}{(b-1+h)^2} \right\}$$

$$+ \lambda_\gamma \left\{ A_2 b\eta\delta v + \frac{bv\cos\gamma}{(b-1+h)} - \frac{b^2 \cos\gamma}{(b-1+h)^2 v} \right\}$$

where λ_h, λ_v, and λ_γ are the costates (Lagrange multipliers) corresponding to the states h, v, and γ respectively. The unconstrained optimal control is obtained from

$$\frac{\partial \mathcal{H}}{\partial \eta} = 0$$

This leads to

$$\eta = E_m \lambda_\gamma / v\lambda_v; \quad \text{where, } E_m = (L/D)_{max}$$

Realistically, the control C_L is bounded by the aerodynamic characteristics of the vehicle. Thus, for the constrained control,

$$|C_L| \le C_{Lmax} \quad \text{or} \quad |\eta| \le c_{max}$$

The costate (adjoint) variables λ's are solved from

$$\frac{d\lambda_h}{d\tau} = -\frac{\partial \mathcal{H}}{\partial h}; \quad \frac{d\lambda_v}{d\tau} = -\frac{\partial \mathcal{H}}{\partial v}; \quad \frac{d\lambda_\gamma}{d\tau} = -\frac{\partial \mathcal{H}}{\partial \gamma} \quad (8)$$

Boundary Conditions

The initial and final boundary conditions are given for the normalized altitude h as

$$h(\tau=0) = 1.0, \quad h(\tau=\tau_f) = 1.0$$

and for the normalized velocity v, and the flight path angle γ as

$$(2-v_e^2)a_d^2 - 2a_d + v_e^2\cos^2\gamma_e = 0$$

$$(2-v_f^2)a_c^2 - 2a_c + v_f^2\cos^2\gamma_f = 0 \quad (9)$$

The above relations result from the considerations of energy conservation and angular momentum conservation applied to the HEO-to-entry elliptic transfer orbit and exit-to-LEO elliptic transfer orbit, respectively. The remaining boundary conditions are obtained from the transversality conditions on the costates. Thus, the optimization procedure, requiring the solution of state equations (7) and the corresponding costate equations (8) along with the boundary conditions (9), leads to the formation of a TPBVP, which is solved by using a multiple shooting method.

3.5 Numerical Data and Results

A typical set of numerical values used for simulation purposes is given below [3].

$$C_{DO} = 0.10; \quad K = 1.11; \quad m/S = 300 \text{ kg/meter}^2$$

$$\rho_s = 1.225 \text{ kg/meter}^3; \quad \mu = 3.986 \times 10^{14} \text{ meter}^3/\text{sec}^2$$

$$\beta = 1/6900 \text{ meter}^{-1}; \quad R_E = 6378 \text{ km}$$

$$H_a = 120 \text{ km}; \quad R_d = 42241 \text{ km}; \quad R_c = 6728 \text{ km}$$

Using the above data, simulations are carried out. Time histories of altitude H, velocity V, flight path angle γ, and lift coefficient η, for a total flight time of 1620 seconds, are shown in Figure 7. The maximum positive lift is used initially to recover from the downward plunge. As the flight path angle becomes positive, the maximum negative lift is used to level off the flight.

Figure 7(a) shows the time history of altitude. The spacecraft enters and exits the atmosphere at an altitude of 120 km. The minimum altitude reached is 50.65 km. The velocity versus time is shown in Figure 7(b). The vehicle enters the atmosphere with a velocity of 10309 m/sec and leaves the atmosphere with a speed of 7880 m/sec, thus giving a velocity reduction of 2429 m/sec. The profile of flight path angle with time is shown in Figure 7(c). The spacecraft enters the atmosphere with an inclination of -6.5428 degrees and exits with +0.5045 degrees. The control history is shown in Figure 7(d). The vehicle enters the atmosphere with maximum lift capability and switches to the minimum lift coefficient and then gradually increases during the remaining flight.

The minimum-fuel transfer requires a deorbit (impulse) characteristic velocity ΔV_d of 1496.07 m/sec and a reorbit characteristic velocity ΔV_c of 87.75 m/sec, with a total characteristic velocity of 1583.82 m/sec. Compare this aeroassisted transfer with the Hohmann transfer, which is maneuvered entirely in outer space, and has a deorbit characteristic velocity ΔV_{dh} of 1461.6 m/sec and the reorbit characteristic velocity ΔV_{ch} of 2412.83 m/sec, giving a total characteristic velocity for Hohmann transfer ΔV_h of 3874.43 m/sec. This shows that the saving due to coplanar, aeroassisted transfer over Hohmann transfer is 59.12 percent. In the case of idealized transfer which follows a grazing trajectory along the atmospheric boundary, the deorbit characteristic velocity ΔV_{di} is 1485.62 m/sec and reorbit characteristic velocity ΔV_{ci} is 67.22 m/sec, thus giving a total characteristic velocity of 1552.84 m/sec. A peak heating rate of 283.21 W/sq.cm. and a peak dynamic pressure is 71.9 kN/sq.m. are observed.

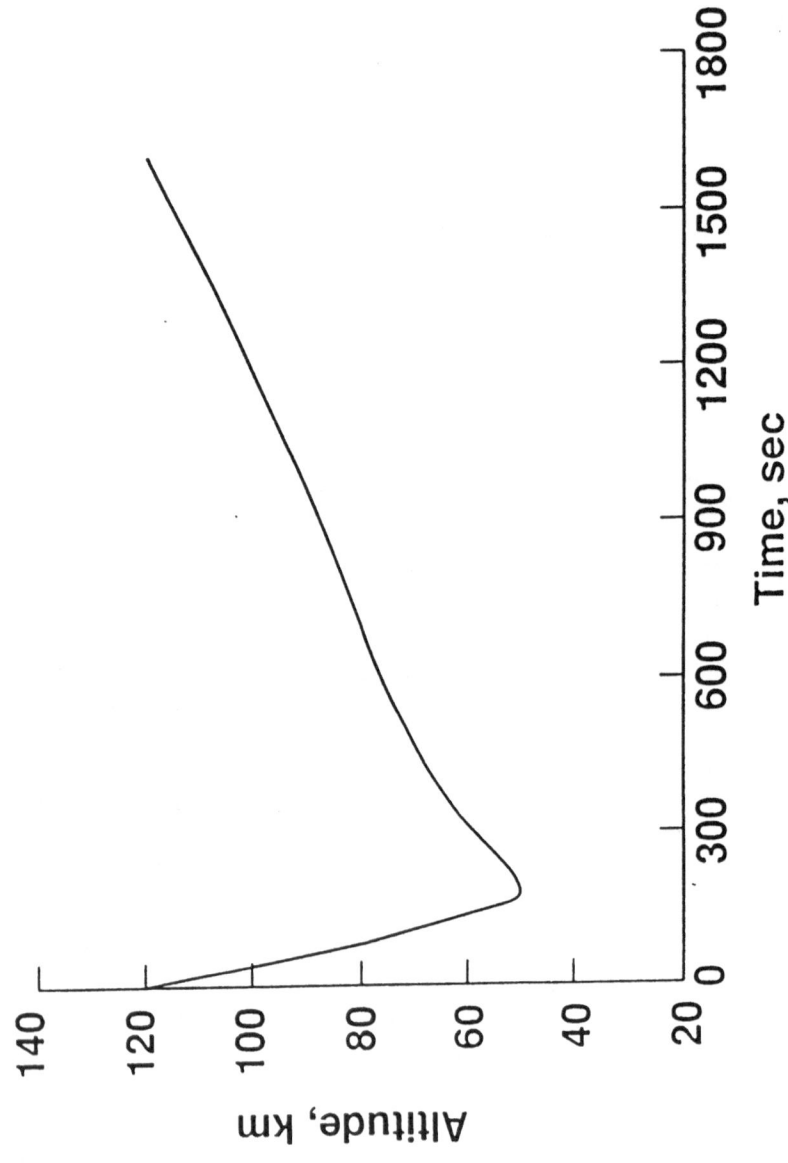

Fig. 7(a)   Time History of Altitude

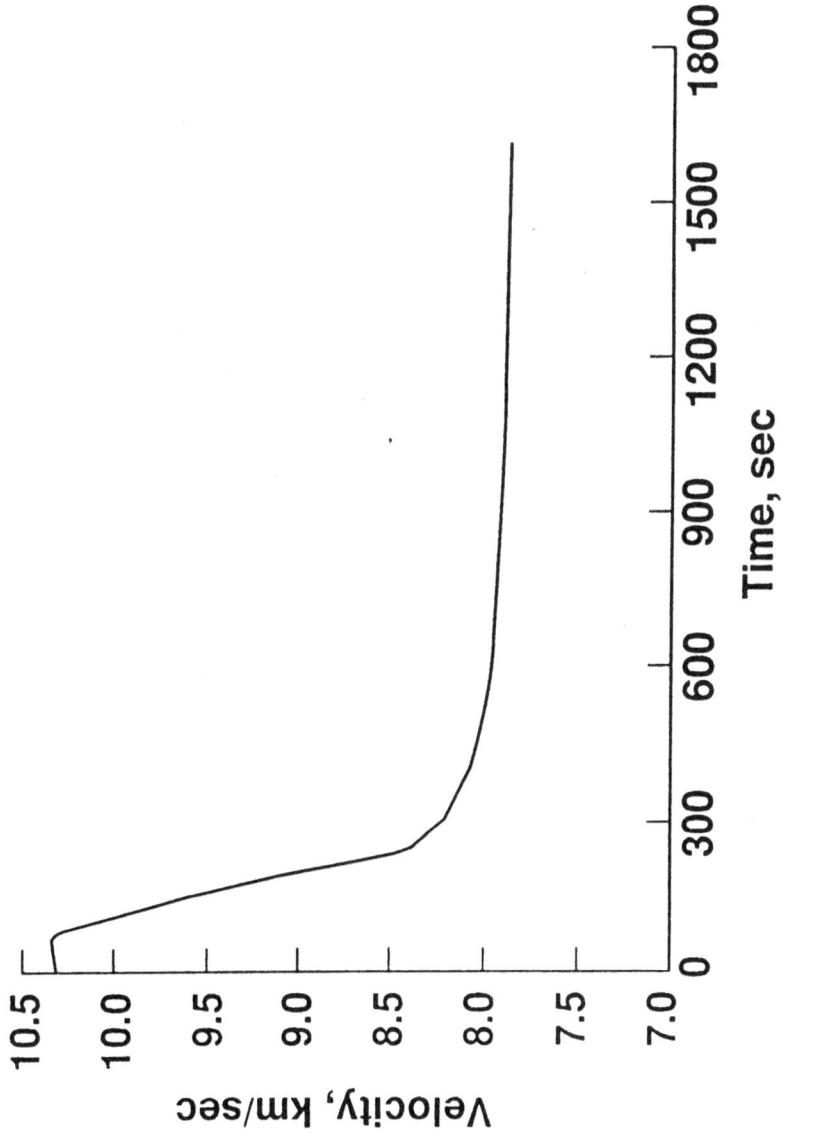

Fig. 7(b) Time History of Velocity

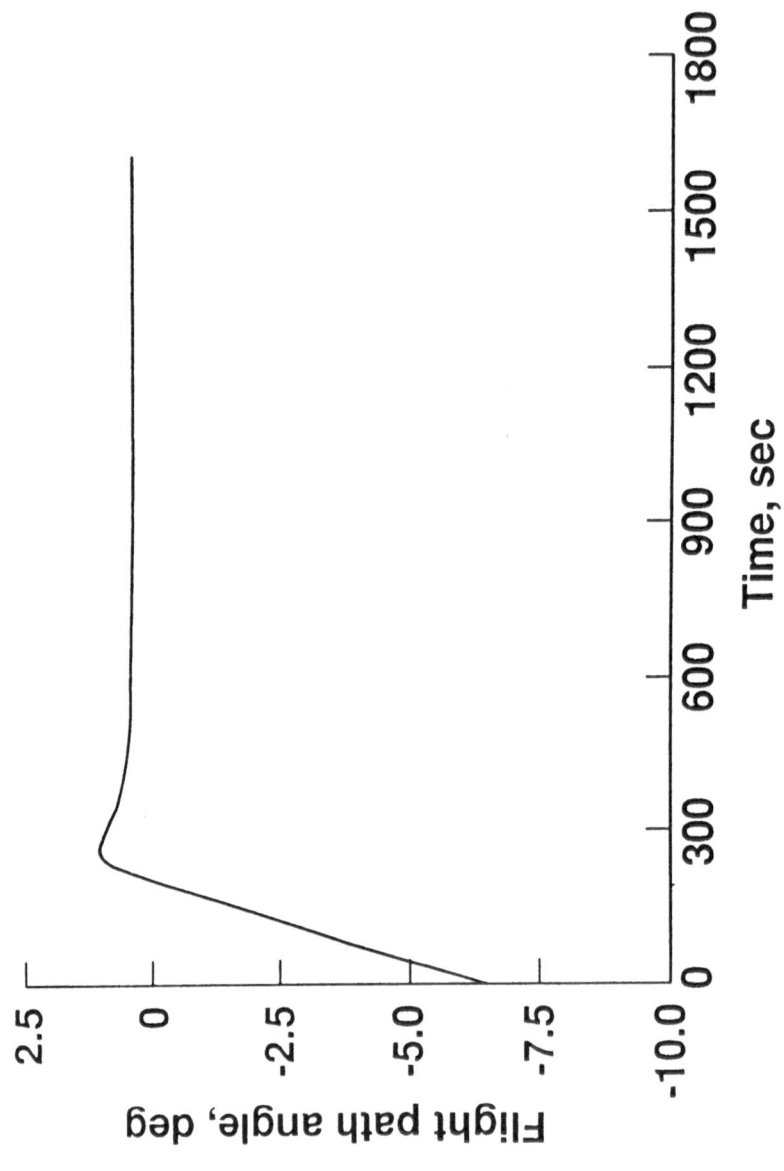

Fig. 7(c)   Time History of Flight Path Angle

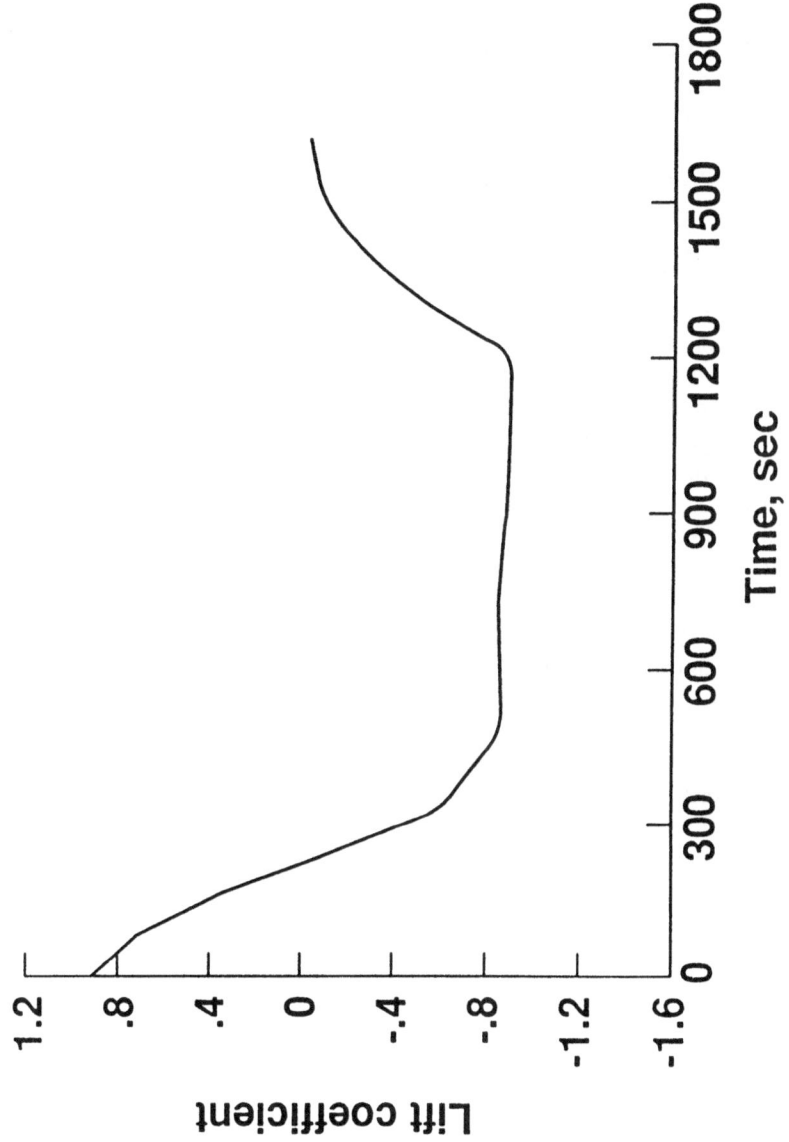

Fig. 7(d)   Time History of Lift Coefficient

Multiple Shooting Method

The determination of optimal control requires the solution of a sixth order, nonlinear TPBVP consisting of state equations (7) and costate equations (8) and the associated boundary conditions (9). This can only be done by numerical methods. The multiple shooting method is one of the powerful methods for solving nonlinear TPBVP's. The corresponding OPTSOL code was developed by Deutsche Forschungs-und Versuchsanstalt fur Luft-und Raumfahrt (DFVLR) at Oberpfaffenhofen, Germany [7,8].

In solving any boundary value problem with the given initial and final conditions, assume additional initial data and integrate forward so that the solution satisfies the given final condition as well. This is called a simple shooting method. Here, the convergence of the solution is highly sensitive to the assumed initial data. It is found that the error due to inaccurate initial data can be made arbitrarily small by performing the integration over sufficiently smaller subdivided panels within the given interval and thereby leading to the multiple shooting method. Thus, the multiple shooting method is a simultaneous application of the simple shooting method at several points within the interval of integration. Here, the trajectory may be restarted at intermediate points using new guesses. Jacobian matrices are formed for each segment. The resulting iteration scheme, based on reducing all discontinuities at internal grid points to zero, leads to a system of linear algebraic equations.

Figure 8 shows the successive approximations of the altitude H, during the course of 0, 9, and 22 iterations. For the sake of clarity only 4 out of 20 intervals are shown. The initial guessed value for the altitude is 120 km at every interval. It can be seen how the initially large jumps at the subdivision points of the multiple shooting method are "flattened out" with the increase of iterations.

3.6 Concluding Remarks

A brief treatment of the three types of orbital transfers, Hohmann transfer, idealized transfer, and realistic transfer, established the fact that the realistic transfer is the upper bound of the idealized (grazing) transfer.

In this Chapter, the minimization of fuel consumption during the atmospheric portion of an aeroassisted, coplanar, orbital transfer vehicle has been addressed. The resulting two-point, boundary value problem was solved by using an efficient multiple shooting algorithm. Simulations have been carried out to obtain fuel optimal trajectories for flying the vehicle through the atmosphere. The strategy for the atmospheric portion of the minimum-fuel transfer is to fly at the maximum lift-to-drag ratio L/D initially in order to recover from the downward plunge, and then to fly at a negative L/D to level off the flight such that the vehicle skips out of the atmosphere with a flight path angle near zero degrees.

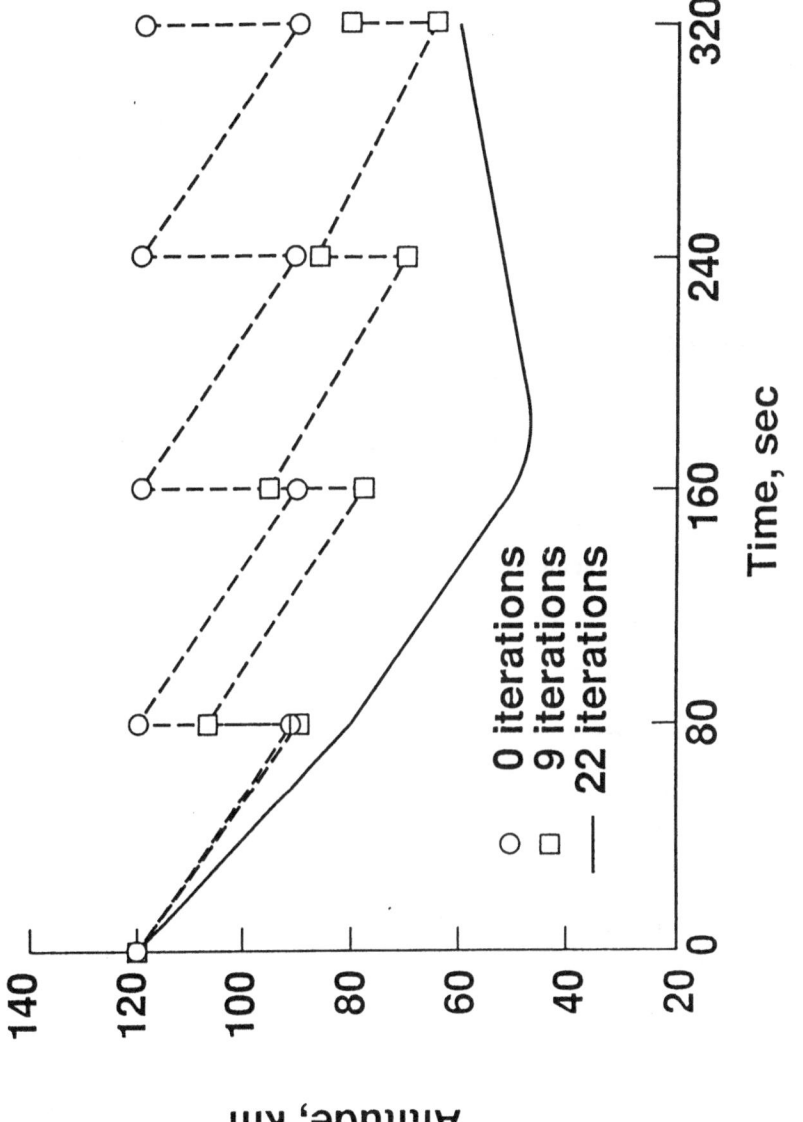

Fig. 8 Successive Approximations for Altitude

Nomenclature

$A \quad = S\rho_s/2m$

$A_1 \quad = C_{DO}S\rho_sH_a/2m$

$A_2 \quad = C_{LR}S\rho_sH_a/2m$

AOTV: Aeroassisted orbital transfer vehicle

$a_d \quad = R_d/R_a$

$a_c \quad = R_c/R_a$

$b \quad = R_a/H_a$

C_D: drag coefficient

C_{DO}:   zero-lift drag coefficient

C_L :   lift coefficient

C_{LR}:   lift coefficient for maximum lift-to-drag ratio

D :   drag force

E_m :   maximum value of L/D

g :   gravitational acceleration

H :   altitude

HEO : high Earth orbit

$\mathscr{H}$:   Hamiltonian

J :   performance index

K :   induced drag factor

L :   lift force

L/D : lift-to-drag ratio

LEO : low Earth orbit

m :   vehicle mass

OTV : orbital transfer vehicle

R :   distance from Earth center to vehicle center of gravity

R_a :   radius of the atmospheric boundary

R_c :   radius of the low Earth orbit

R_d :   radius of the high Earth orbit

R_E : radius of Earth

R_s : distance from vehicle center of gravity to surface level

S :   aerodynamic reference area

SSO : space station orbit

t :   time

V :   velocity

v :   normalized velocity

β :   inverse atmospheric scale height

γ : flight path angle
δ : normalized density
λ : costate (Lagrange) variable
μ : gravitational constant of Earth

$\eta = C_L/C_{LR}$

ρ : density
τ : normalized time
ΔV : characteristic velocity
Δv : normalized characteristic velocity

subscripts

c : circularization or reorbit
d : deorbit
e : entry to atmosphere
f : exit from atmosphere
h : Hohmann transfer
i : idealized transfer
s : surface level

References

[1] *Pioneering the Space Frontier*, The Report of the National Space Commission on Space, Banton Books Inc., New York, May 1986.

[2] G. D. Walberg, "A survey of aeroassisted orbital transfer," *J. Spacecraft*, 22, 3-18, 1985.

[3] K. D Mease, and N. X. Vinh, "Minimum-fuel aeroassisted coplanar orbit transfer using lift modulation," *J. Guidance, Control, and Dynamics*, 8, 134-141, 1985.

[4] A. Miele, V. K. Basapur, and W. Y. Lee, "Optimal trajectories for aeroassisted coplanar orbit transfer," *J. Opt. Theory & Appl.*, 52, 1-24, 1987.

[5] M. H. Kaplan, *Modern Spacecraft Dynamics and Control*, John Wiley & Sons, New York, 1976.

[6] J. P. Marec, *Optimal Space Trajectories*, Elsevier Scientific Publishing Company, Amsterdam, 1979.

[7] J. Stoer, and R. Bulirsch, *Introduction to Numerical Analysis*, Springer-Verlag, New York, 1980.

[8] H. J. Pesch, "Numerical computation of neighboring optimum feedback control schemes in real-time," *Appl. Math. Optim.*, 5, 231-252, 1979.

****

CHAPTER 4

FUEL OPTIMAL CONTROL FOR
NONCOPLANAR ORBITAL TRANSFER

4.1 Introduction

The main function of space transportation system is to deliver payloads from Earth to various locations in space. Until now, this function has been performed by various rockets, the space shuttle, and expendable upper stages using solid or liquid propellants. In particular, considering the economic benefits and reusability, an orbital transfer vehicle (OTV) is proposed for transporting payloads between low Earth orbit (LEO) and high Earth orbit (HEO). The two basic operating modes contemplated for OTV are a ground-based OTV which returns to Earth after each mission and a space-based OTV which operates out of an orbiting hanger located at the proposed Space Station Freedom.

In a typical mission, a space-based OTV, which is initially at the space station orbit (SSO), is required to transfer a payload to geosynchronous Earth orbit (GEO), pick up another payload, say a faulty satellite, and return to mate with the orbiting hanger at SSO for refurbishment and redeployment of the payload. The OTV on its return journey from GEO to SSO needs to dissipate some of its orbital energy. This can be accomplished by using an entirely propulsive (Hohmann) transfer in space only or a combination of propulsive transfer in space and aeroassisted maneuver in the atmosphere. It has been established that a significant fuel savings and hence increased payload capabilities can be achieved with propulsive and aeroassisted maneuvers instead of all-propulsive maneuvers [1]. This leads to an aeroassisted orbital transfer vehicle (AOTV), which on its return leg of the mission, dips into the Earth's atmosphere, utilizes atmospheric drag to reduce the orbital velocity and employs lift and bank angle modulations to achieve a desired orbital inclination. Basically, the AOTV performs a synergetic maneuver, employing a hybrid combination of

This Chapter is based on D. S. Naidu, "Fuel-optimal trajectories of aeroassisted orbital transfer with plane change," IEEE Trans. Aerospace and Electronic Systems, Vol. 27, pp. 361-369, March 1991. @ 1991 IEEE. Permission given by IEEE is hereby acknowledged.

propulsive maneuver in space and aerodynamic maneuver in the atmosphere. It
is believed that the concept of aeroassisted orbital transfer opens new
mission opportunities for the space transportation system, especially with
regard to the establishment of the permanent space station. The fuel
optimization is an important aspect of orbital transfer missions [2-7].

In Chapter 3, the fuel-optimization problem associated with the
aeroassisted coplanar orbital transfer using multiple shooting method has
been presented [8-10]. The coplanar case is a simpler one, where the only
control is by aerodynamic lift modulation. The vehicle is maneuvered to the
LEO in the same plane as that of the HEO. On the other hand, the present
Chapter focuses on the fuel minimization for the more important noncoplanar
orbital transfer using the multiple shooting method. Here, a desired plane
change is achieved by a combination of lift and bank angle modulations so as
to bring the vehicle to the required orbital plane at LEO to rendezvous with
the orbiting space station.

This Chapter, addresses the fuel-optimal control problem arising in
noncoplanar orbital transfer employing aeroassist technology. The maneuver
involves the transfer from HEO to LEO with a plane change and at the same
time minimization of the fuel consumption. It is known that the change in
velocity, also called the characteristic velocity, is a convenient measure
of fuel consumption. For the minimum-fuel maneuver, the objective is then to
minimize the total characteristic velocity for deorbit, boost, and reorbit
(or   circularization)   for   a   specified   change   in   inclination   angle.
Application of Pontryagin minimum principle leads to a nonlinear,
two-point, boundary value problem (TPBVP), which is solved by using multiple
shooting method [8-10].

4.2 Basic Equations

For the orbital transfer problem, the following assumptions are made.
(i) The initial HEO and final LEO orbits are circular. (ii) The mission is
comprised of three impulses. (iii) The vehicle is represented as a constant
point mass during atmospheric pass. (iv) A Newtonian inverse square
gravitational field is used. (v) Earth's rotation is neglected. (vi) The
atmosphere is exponential. (vii) The vehicle has a parabolic drag polar.

The complete mission from HEO to LEO with atmospheric pass is depicted
in Fig. 1. It is composed of three impulses: first, a deorbit impulse ΔV_d at

HEO to inject the vehicle into a HEO-entry elliptic orbit, second, a boost
impulse ΔV_b at the exit from the atmosphere for the vehicle to attain

sufficient velocity to travel along an exit-LEO elliptic orbit, and finally,
a circularizing impulse ΔV_c to circularize the path of the vehicle. Consider

the basic equations of motion for different phases of deorbit, aeroassist
(or atmospheric flight), boost and reorbit (or circularization).

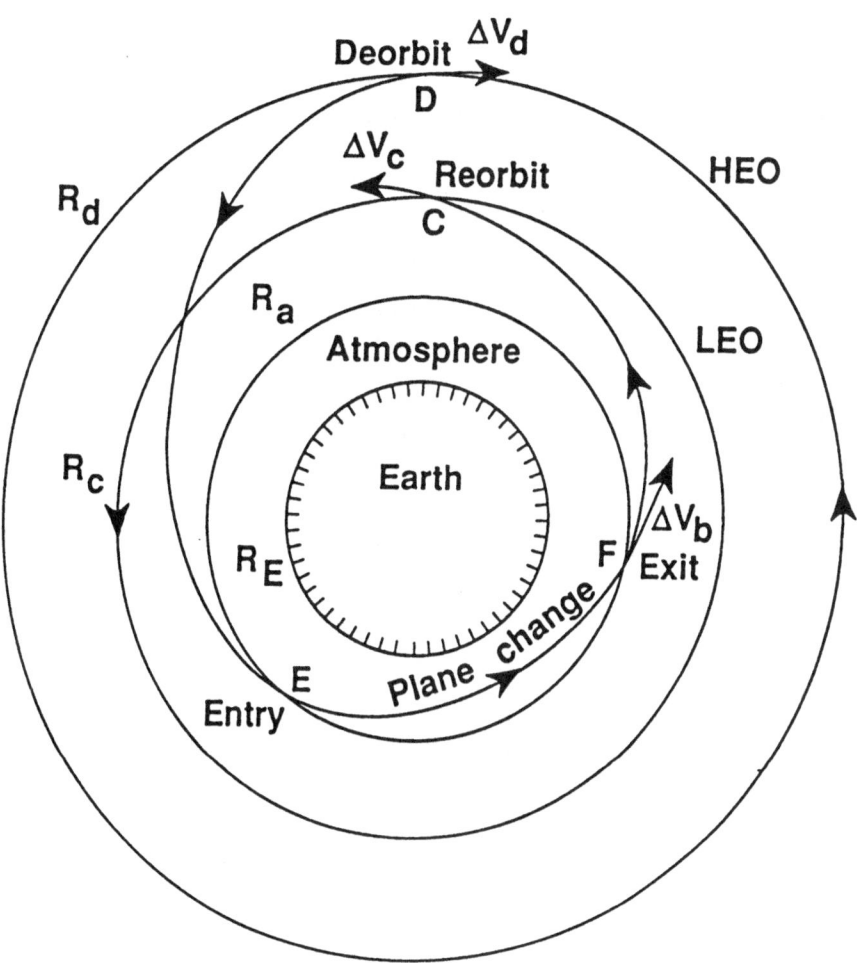

Fig. 1   Aeroassisted Orbital Plane Change

4.2.1 Deorbit Flight

Initially, assume that the spacecraft is in a circular orbit of radius R_d, well outside the Earth's atmosphere, moving with a circular velocity V_d $= \sqrt{\mu/R_d}$. Deorbit is performed by means of an impulse ΔV_d, to transfer the vehicle from the circular orbit to elliptic orbit with perigee low enough to intersect the dense part of the atmosphere [Fig. 1]. Since the elliptic velocity at D is less than the circular velocity at D, the impulse ΔV_d is executed so as to oppose the circular velocity V_d. The deorbit impulse ΔV_d causes the vehicle to enter the atmosphere of radius R_a with a velocity V_e and flight path angle γ_e. It is known that the optimal-energy loss maneuver from the circular orbit is simply the Hohmann transfer and the impulse is parallel and opposite to the instantaneous velocity vector.

Using the principle of conservation of energy and angular momentum at the deorbit point D, and the atmospheric entry point E yields [11],

$$V_e^2/2 - \mu/R_a = (V_d - \Delta V_d)^2/2 - \mu/R_d$$

$$R_a V_e \cos(-\gamma_e) = R_d(V_d - \Delta V_d)$$

from which solving for ΔV_d, gives

$$\Delta V_d = \sqrt{\mu/R_d} - \sqrt{2\mu(1/R_a - 1/R_d)/\left[(R_d/R_a)^2/\cos^2\gamma_e - 1\right]}$$

It is easily seen that the minimum value of the deorbit impulse ΔV_{dm} obtained at $\gamma_e = 0$, corresponds to an ideal transfer wherein the space vehicle grazes along the atmospheric boundary. To ensure proper atmospheric entry, the deorbit impulse ΔV_d must be higher than the minimum deorbit impulse ΔV_{dm} which is given by

$$\Delta V_{dm} = \sqrt{\mu/R_d} - \sqrt{2\mu(1/R_a - 1/R_d)/\left[(R_d/R_a)^2 - 1\right]}$$

4.2.2 Aeroassist (Atmospheric) Flight

During the aeroassist (or atmospheric) flight, the vehicle performs a three-dimensional skip maneuver and using aerodynamic lift and bank angle achieves the necessary the plane change. In this process, the vehicle decelerates due to the atmospheric drag.

The equations of motion for the vehicle during the atmospheric pass are given below [Fig. 2]. The kinematic equations are [2],

$$\frac{dR}{dt} = V\sin\gamma$$

$$\frac{d\theta}{dt} = V\cos\gamma\cos\psi/R\cos\phi$$

$$\frac{d\phi}{dt} = V\cos\gamma\sin\psi/R$$

The force equations are

$$m\frac{dV}{dt} = - D - mg\sin\gamma$$

$$mV\frac{d\gamma}{dt} = L\cos\sigma + m(V^2/R - g)\cos\gamma$$

$$mV\frac{d\psi}{dt} = L\sin\sigma/\cos\gamma - (mV^2/R)\cos\gamma\cos\psi\tan\phi \qquad (1)$$

where,

$$L = C_L\rho SV^2/2; \quad D = C_D\rho SV^2/2; \quad C_D = C_{DO} + KC_L^2$$

$$g = \mu/R^2; \quad R = H + R_E; \quad \rho = \rho_s\exp(-H\beta)$$

Using the normalized variables,

$$\tau = t/\sqrt{R_a^3/\mu} \; ; \quad v = V/\sqrt{\mu/R_a}$$

and the dimensionless constants,

$$h = H/H_a; \quad b = R_a/H_a; \quad \delta = \rho/\rho_s = \exp(-h\beta H_a)$$

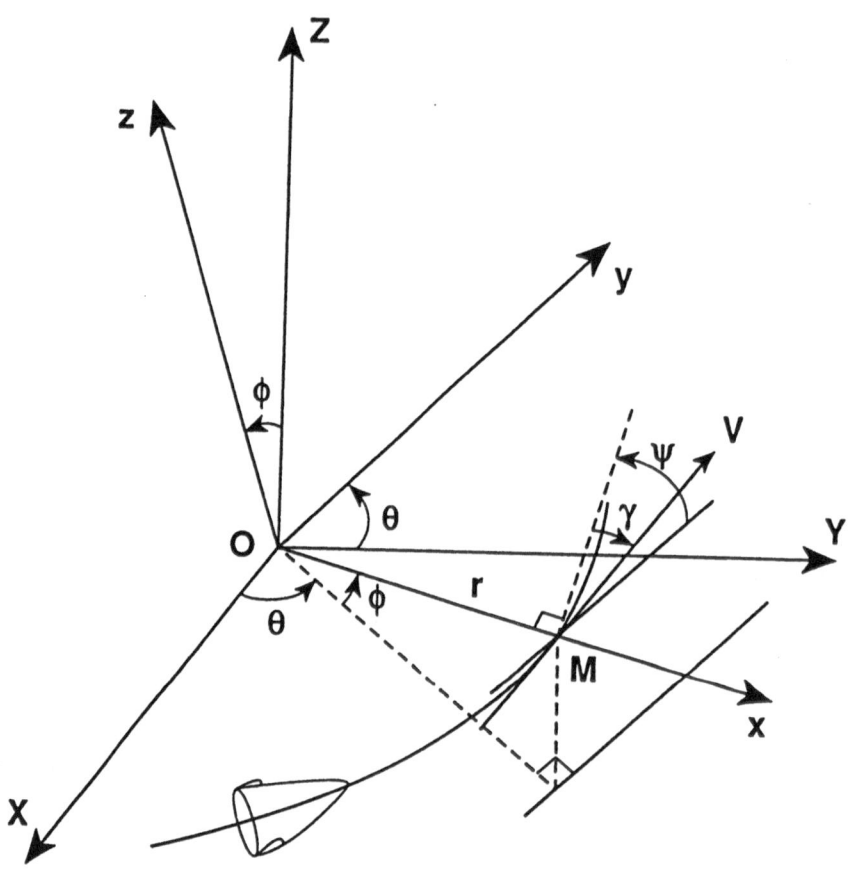

Fig. 2 Coordinate System

$$\eta = C_L/C_{LR}; \quad C_{LR} = \sqrt{C_{DO}/K}$$

in (1), we get the normalized form as

$$\frac{dh}{d\tau} = bv\sin\gamma$$

$$\frac{d\theta}{d\tau} = \frac{bv\cos\gamma\cos\psi}{(b-1+h)\cos\phi}$$

$$\frac{d\phi}{d\tau} = \frac{bv\cos\gamma\sin\psi}{(b-1+h)}$$

$$\frac{dv}{d\tau} = -A_1 b(1+\eta^2)\delta v^2 - \frac{b^2\sin\gamma}{(b-1+h)^2}$$

$$\frac{d\gamma}{d\tau} = A_2 b\eta\delta v\cos\sigma + \frac{bv\cos\gamma}{(b-1+h)} - \frac{b^2\cos\gamma}{(b-1+h)^2 v}$$

$$\frac{d\psi}{d\tau} = \frac{A_2 b\delta\eta v\sin\sigma}{\cos\gamma} - \frac{bv\cos\gamma\cos\psi\tan\phi}{(b-1+h)} \qquad (2)$$

where, $A_1 = C_{DO}S\rho_s H_a/2m$; $A_2 = C_{LR}S\rho_s H_a/2m$

From the above equations of motion, see clearly that during the atmospheric maneuver, if the lift vector L is rotated about the velocity vector V through the bank angle σ, it creates a lateral force component $L\sin\sigma$ orthogonal to the vertical plane that has the effect of changing the heading angle ψ. At the end of the maneuver, the vehicle is already in vacuum and hence there is no lift. The equations (1) for the cross range angle ϕ, and the heading angle ψ, become [2],

$$\frac{d\phi/dt}{d\psi/dt} = -\frac{\tan\psi}{\tan\phi}$$

integration of which yields,

$$\cos\phi\cos\psi = \cos i$$

where, i is the orbital inclination. For small values of cross range angle ϕ, the orbital inclination i is given by the heading angle ψ itself. Thus, the total change in the heading corresponds to the change in orbital inclination (plane change).

4.2.3 Boost and Reorbit Flight

During the atmospheric flight, the vehicle performs the desired plane change and dissipates some energy due to atmospheric drag. Therefore, a second impulse is required to boost the vehicle back to orbital altitude. The vehicle exits the atmosphere at point F, with a velocity V_f and flight path angle γ_f. The additional impulse ΔV_b, required at the exit point F for boosting into an elliptic orbit with apogee radius R_c and the reorbit impulse ΔV_c required to insert the vehicle into a circular orbit at point C, are obtained by using the principle of conservation of energy and angular momentum at the exit point F, and the circularization point C. This leads to [11],

$$(V_f + \Delta V_b)^2/2 - \mu/R_a = (V_c - \Delta V_c)^2/2 - \mu/R_c$$

$$(V_f + \Delta V_b)R_a \cos\gamma_f = R_c(V_c - \Delta V_c)$$

Solving for ΔV_b and ΔV_c,

$$\Delta V_b = \sqrt{2\mu(1/R_a - 1/R_c)/\left[1 - (R_a/R_c)^2\cos^2\gamma_f\right]} - V_f$$

$$\Delta V_c = \sqrt{\mu/R_c} - \sqrt{2\mu(1/R_a - 1/R_c)/\left[(R_c/R_a)^2/\cos^2\gamma_f - 1\right]} \tag{3}$$

Finally, the vehicle is in a circular orbit (of radius R_c) moving with the velocity $V_c = \sqrt{\mu/R_c}$.

4.3 Fuel Optimal Control

For the minimum-fuel maneuver, the objective is then to minimize the total characteristic velocity for a specified change in heading angle. A convenient performance index is the sum of the characteristic velocities for deorbit, boost, and reorbit. Thus,

$$J = \Delta V_d + \Delta V_b + \Delta V_c$$

where, ΔV_d, ΔV_b, and ΔV_c are the deorbit, boost, and reorbit characteristic velocities respectively, and are obtained as

$$\Delta V_d = \sqrt{\mu/R_d} - (R_a/R_d)V_e \cos(-\gamma_e)$$

$$\Delta V_c = \sqrt{\mu/R_c} - (R_a/R_c)(V_f + \Delta V_b)\cos\gamma_f \qquad (4)$$

In the normalized form, the performance index becomes,

$$J = \Delta v = \Delta v_d + \Delta v_b + \Delta v_c$$

where,

$$\Delta v_d = \sqrt{1/a_d} - (v_e/a_d)\cos(-\gamma_e)$$

$$\Delta v_c = \sqrt{1/a_c} - \left[(v_f + \Delta v_b)/a_c\right]\cos\gamma_f$$

$$a_d = R_d/R_a; \quad a_c = R_c/R_a; \quad \Delta v_d = \Delta V_d/\sqrt{\mu/R_a}$$

$$\Delta v_c = \Delta V_c/\sqrt{\mu/R_a}$$

Note that for a given circular orbit of radius R_c, the impulses ΔV_b and ΔV_c are completely determined by the velocity V_f and the flight path angle γ_f at the atmospheric exit. The velocity V and the flight path angle γ_e at the entry point are dependent only on the magnitude of the deorbit impulse ΔV_d. For a specified atmospheric entry (i.e., for a given perigee altitude occurring within the atmosphere), we have a fixed value of ΔV_d and hence fixed values of entry velocity V_e, and entry flight path angle γ_e. Therefore, the optimal control problem for the minimum fuel consumption is confined only to the segment of the trajectory within the atmosphere. Hence, the performance index is more appropriately written as

$$J = \Delta V_d + \Delta V_b(V_f, \gamma_f) + \Delta V_c(V_f, \gamma_f)$$

Ideally, as seen from (3), the minimum value of boost impulse ΔV_b is zero, when the exit velocity V_f is made equal to the perigee velocity of the exit-LEO elliptic orbit. Also, the minimum value of reorbit impulse ΔV_c is obtained when the vehicle exits with zero flight path inclination γ_f.

The first step in the optimization procedure using Pontryagin principle is to formulate the Hamiltonian as [2],

$$\mathcal{H} = \lambda_h \, bv\sin\gamma + \lambda_v \left\{ - A_1 b(1+\eta^2)\delta v^2 - \frac{b^2 \sin\gamma}{(b-1+h)^2} \right\}$$

$$+ \lambda_\gamma \left\{ A_2 b\eta\delta v\cos\sigma + \frac{bv\cos\gamma}{(b-1+h)} - \frac{b^2\cos\gamma}{(b-1+h)^2 v} \right\} + \lambda_\phi \left\{ \frac{bv\cos\gamma\sin\psi}{(b-1+h)} \right\}$$

$$+ \lambda_\psi \left\{ \frac{A_2 b\delta\eta v\sin\sigma}{\cos\gamma} - \frac{bv\cos\gamma\cos\psi\tan\phi}{(b-1+h)} \right\}$$

where λ's are the costates corresponding to the states. The down range angle θ does not enter the right hand side of (1) and hence need not be considered for the optimization process.

The optimal control equations for lift and bank angle are given by

$$\frac{\partial \mathcal{H}}{\partial \eta} = 0; \quad \frac{\partial \mathcal{H}}{\partial \sigma} = 0$$

leading to

$$\eta = C_{LR}\omega/C_{D0} 2v\lambda_v; \quad \tan\sigma = \lambda_\psi/\lambda_\gamma \cos\gamma \qquad (5)$$

where

$$\omega = \sqrt{\lambda_\gamma^2 + \left(\lambda_\psi/\cos\gamma\right)^2}$$

The control C_L is bounded by the aerodynamic characteristics of the vehicle. Thus, for the constrained control,

$$|C_L| \le C_{Lmax} \quad \text{or} \quad |\eta| \le c_{max}$$

The costate (adjoint) equations are given by

$$\frac{d\lambda_h}{d\tau} = -\frac{\partial \mathcal{H}}{\partial h}, \quad \frac{d\lambda_v}{d\tau} = -\frac{\partial \mathcal{H}}{\partial v}, \quad \frac{d\lambda_\gamma}{d\tau} = -\frac{\partial \mathcal{H}}{\partial \gamma}$$

$$\frac{d\lambda_\phi}{d\tau} = -\frac{d\mathcal{H}}{d\phi}, \quad \frac{d\lambda_\psi}{d\tau} = -\frac{d\mathcal{H}}{d\psi}$$

4.3.1 Boundary Conditions

The initial and final boundary conditions are given for the normalized altitude h as

$$h(\tau=0) = 1.0, \quad h(\tau=\tau_f) = 1.0$$

and for the normalized velocity v, and the flight path angle γ as

$$(2-v_e^2)a_d^2 - 2a_d + v_e^2\cos^2\gamma_e = 0$$

$$\left[2-(v_f+\Delta v_b)^2\right]a_c^2 - 2a_c + (v_f+\Delta v_b)^2\cos^2\gamma_f = 0 \tag{6}$$

The equations are obtained by eliminating ΔV_d and ΔV_c from the original equations for the characteristic velocities. The remaining multiplier boundary conditions are obtained from the transversality conditions on the costates. Thus, the optimization procedure, requiring the solution of the state and the costate equations along with the boundary conditions leads to a nonlinear TPBVP, which can only be solved by numerical methods.

4.3.2 Multiple Shooting Method

The multiple shooting method is a powerful method for solving nonlinear TPBVP [8-10]. In solving any boundary value problem with the given initial and final conditions, a simple shooting method is used. Here assume additional initial data and integrate forward so that the solution satisfies the given final condition. However, the convergence of the solution is very much dependent on the assumed initial data. In multiple shooting method, the error due to inaccurate initial data can be made arbitrarily small by performing the integration over sufficiently smaller subdivided intervals. Thus, the multiple shooting method can be thought of as a simultaneous application of the simple shooting method at several points within the interval of integration. Here, the trajectory may be restarted at intermediate points using new guesses. The resulting iteration scheme, based on reducing all discontinuities at internal grid points to zero, leads to a system of linear algebraic equations. The corresponding OPTSOL code, developed by DFVLR at Oberpfaffenhofen, Germany, was used for solving the present problem.

4.4 Typical Data and Results

A typical AOTV configuration [12] with L/D of about 1.5 is shown in Fig. 3. The liquid oxygen is stored in two separate tanks to provide a tapered nose, and inflated chins are used to continue this tapering along the body. A large deployable flap is provided to trim the vehicle at low angles of attack for maximum L/D performance. A representative set of numerical values used for a complete mission from GEO to SSO at an altitude of 556 km is given below [3,6].

$$C_{D0} = 0.1; \quad K = 1.11; \quad m/S = 300 \text{ kg/m}^2$$

$$\rho_s = 1.225 \text{ kg/m}^3; \quad \mu = 3.96772 \times 10^{14} \text{ m}^3/\text{sec}^2$$

$$\beta = 1/6900 \text{ m}^{-1}; \quad R_E = 6356.766 \text{ km}$$

$$H_a = 120 \text{ km}; \quad R_d = 42240.766 \text{ km}; \quad R_c = 6912.766 \text{ km}$$

Using the above mentioned data, the optimal solution has the following entry and exit status.

Entry status:

$$H_e = 120 \text{ km}; \quad V_e = 10305.58 \text{ m/sec}$$

$$\gamma_e = -6.0 \text{ degrees}; \quad \phi_e = 0; \quad \psi_e = 0$$

Exit status:

$$H_f = 120 \text{ km}; \quad V_f = 7462.35 \text{ m/sec}$$

$$\gamma_f = 0.1595 \text{ deg}; \quad \phi_f = 11.95 \text{ deg}$$

$$\psi_f = 24.1 \text{ deg}; \quad \text{total flight time} = 520 \text{ sec}$$

Characteristic velocities:

Deorbit charact. velocity, ΔV_d = 1493.32 m/sec

Boost charact. velocity, ΔV_b = 490.77 m/sec

Reorbit charact. velocity, ΔV_c = 124.61 m/sec

Total charact. velocity ΔV = 2108.7 m/sec

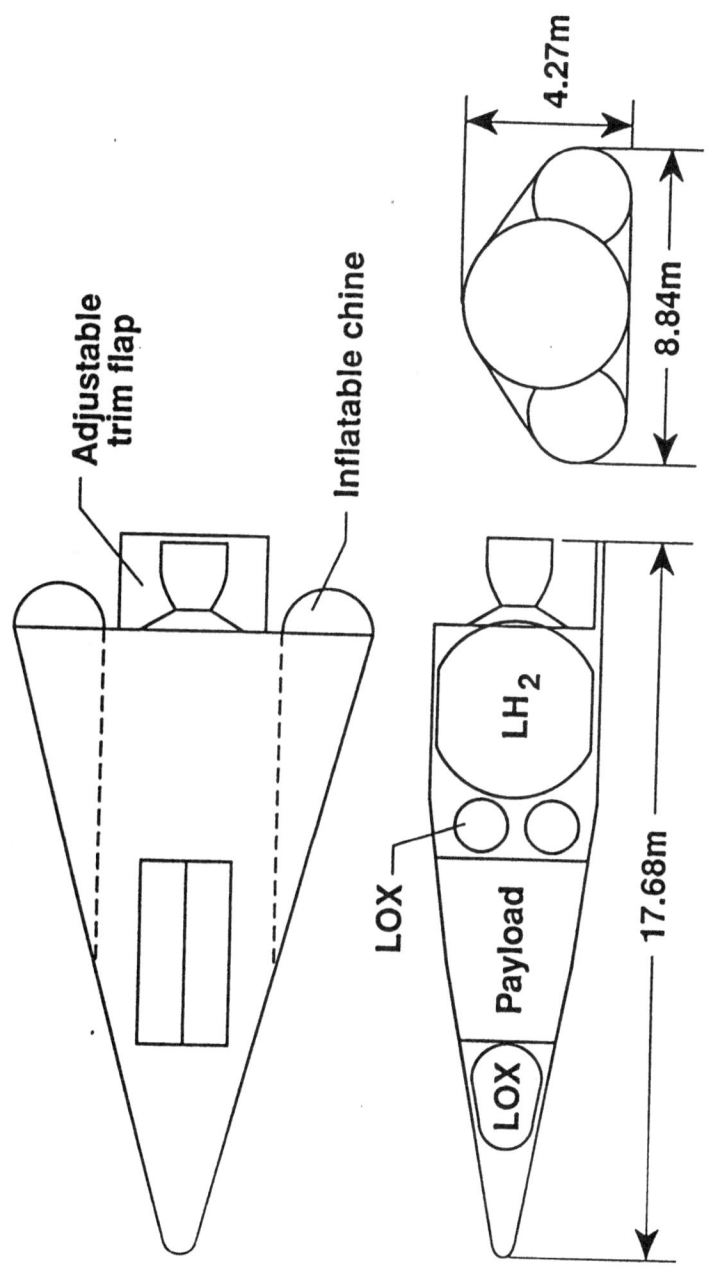

Fig. 3 Vehicle Configuration

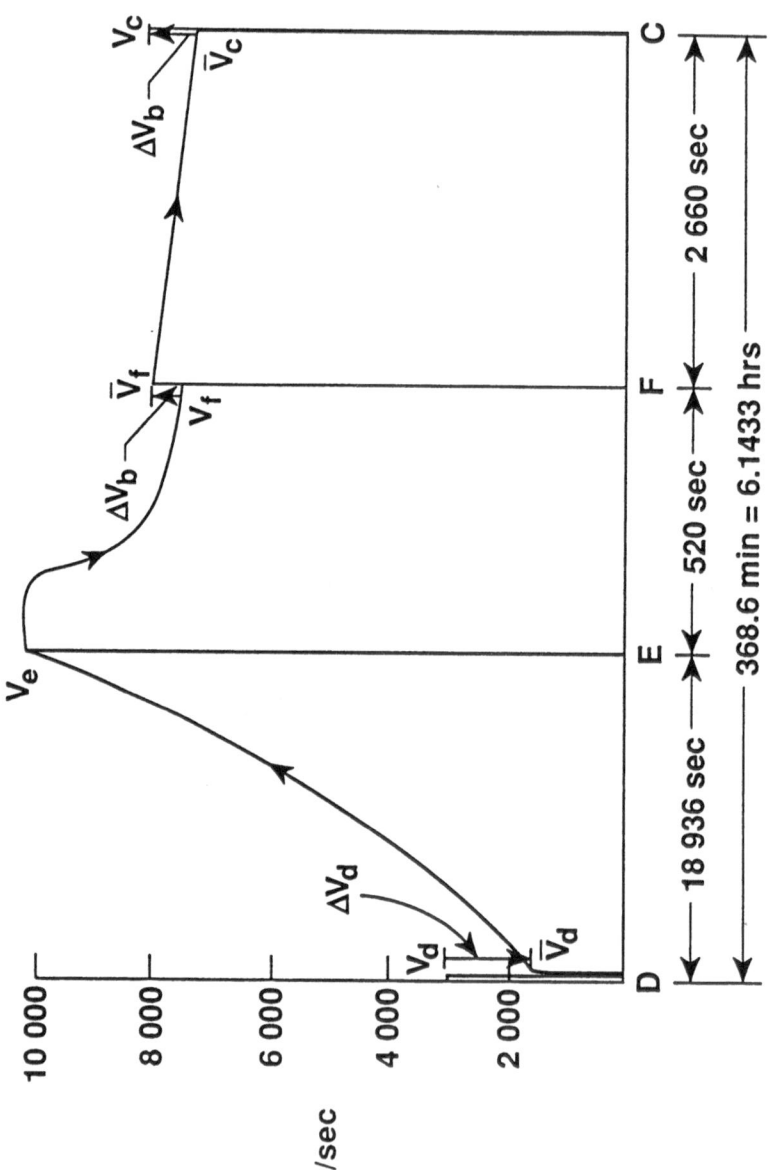

Fig. 4 Velocity Profile for Complete Mission

The complete mission in terms of the velocity profile is shown in Fig. 4. Initially, the vehicle is in a circular orbit at GEO moving at a speed V_d = 3064.82 m/sec. A deorbit impulse ΔV_d = 1493.32 m/sec is executed to fly the vehicle along the GEO-entry elliptic orbit. The elliptic velocity at the deorbit point D is $\bar{V}_d = V_d - \Delta V_d$ = 1571.5 m/sec. At the atmospheric interface E of altitude H_a = 120 km, the vehicle attains an orbital velocity V_e = 10305.58 m/sec. During the atmospheric maneuver, the velocity of the vehicle is depleted and the exit velocity is V_f = 7462.35 m/sec. In order to attain the desired SSO altitude H_c = 556 km, a boost impulse ΔV_b = 490.77 m/sec is required at the exit F from the atmosphere. Then the elliptic velocity at the exit is $\bar{V}_f = V_f + \Delta V_b$ = 7953.12 m/sec. The vehicle travels then along the exit-SSO elliptic path and has a velocity $\bar{V}_c$ = 7451.47 m/sec at the reorbit point C. In order to insert the vehicle into a circular orbit at this altitude H_c = 556 km, a reorbit impulse ΔV_c = 124.61 m/sec is imparted. The vehicle is now in a circular orbit at SSO moving with a speed of $V_c = \bar{V}_c + \Delta V_c$ = 7576.08 m/sec.

Fig. 5(a) shows the time history of altitude. The spacecraft enters and exits the atmosphere at the altitude of 120 km. The minimum altitude reached is 44.72 km. Fig. 5(b) shows a velocity reduction of 2843.23 m/sec. The profile of flight path angle with time is shown in Fig. 5(c). The spacecraft enters the atmosphere with an inclination of -6.00 degrees and exits with 0.1595 degrees. The time history of cross range ϕ is shown in Fig. 5(d) which has a value of 11.95 degrees at the end of the atmospheric maneuver. Fig. 6(e) shows the variation of heading angle ψ, which shows that the atmospheric maneuver provides an orbital inclination of 24.1 degrees.

The control history is shown in Fig. 6(a). The vehicle enters the atmosphere with maximum lift capability and decreases slowly during the remaining flight. Fig. 6(b) shows the variation of bank angle during the atmospheric flight. Initially the vehicle enters the atmosphere with a bank angle of 144.5 degrees to pull the vehicle into the atmosphere but slowly drops to about 75 degrees and maintain at a value of 96 degrees for most of the remainder of the flight. Fig. 7(a) shows the peak heating rate of 402.64 W/sq. cm. As shown in Fig. 7(b), the peak dynamic pressure is 80.73 KN/sq. m.

Fig. 8 shows the successive approximations of the altitude H, during the course of 0, 15, and 30 iterations in using the multiple shooting method. For the sake of clarity only 4 out of 20 intervals are shown. The initial guessed value for the altitude is 120 km at every interval. It can be seen how the initially large jumps at the subdivision points are "flattened out" with the increase of iterations.

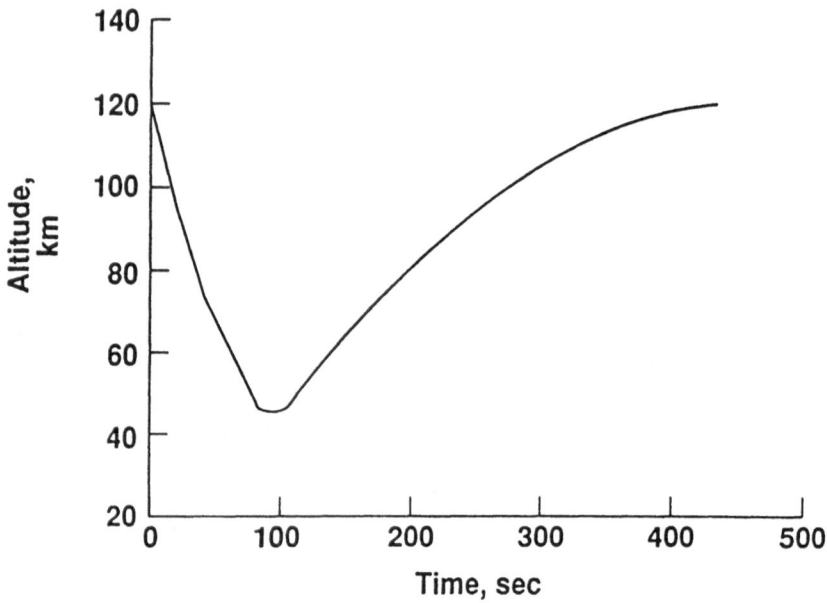

Fig. 5(a)   Time History of Altitude

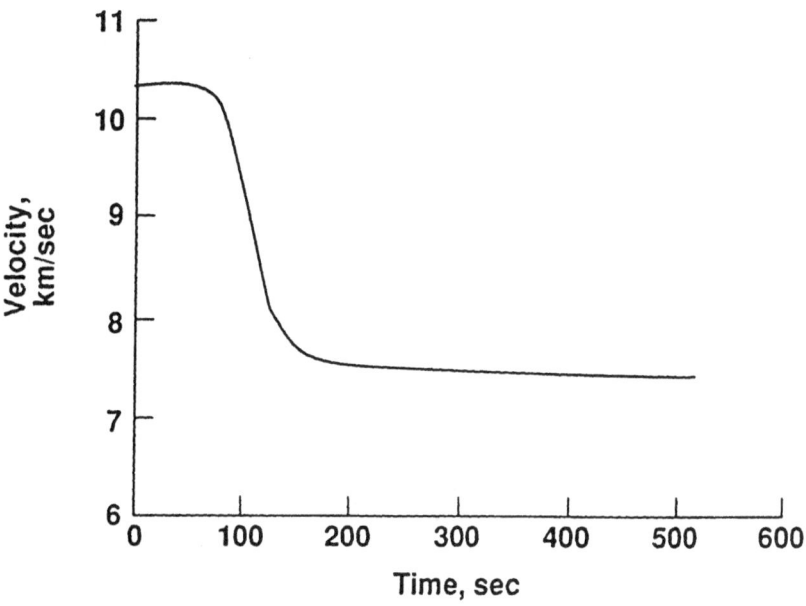

Fig. 5(b) Time History of Velocity

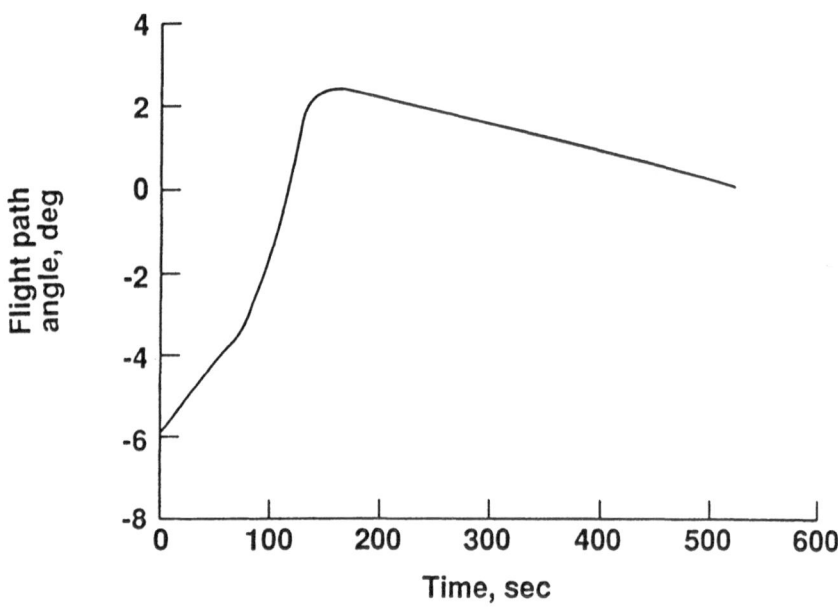

Fig. 5(c) Time History of Flight Path Angle

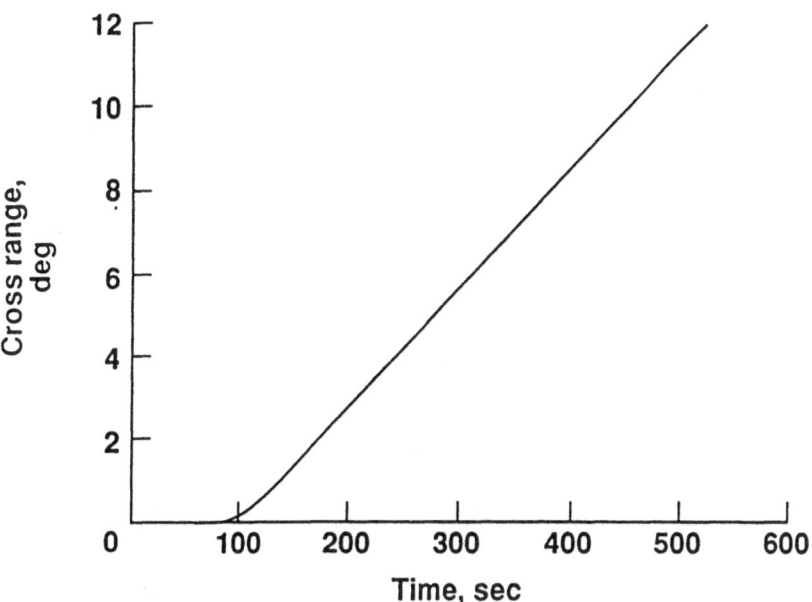

Fig. 5(d)   Time History of Cross Range Angle

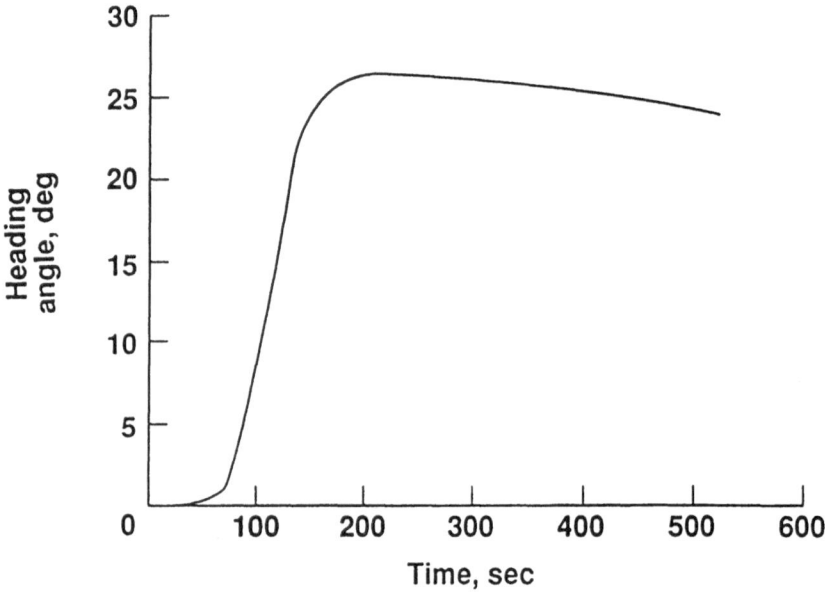

Fig. 5(e)   Time History of Heading Angle

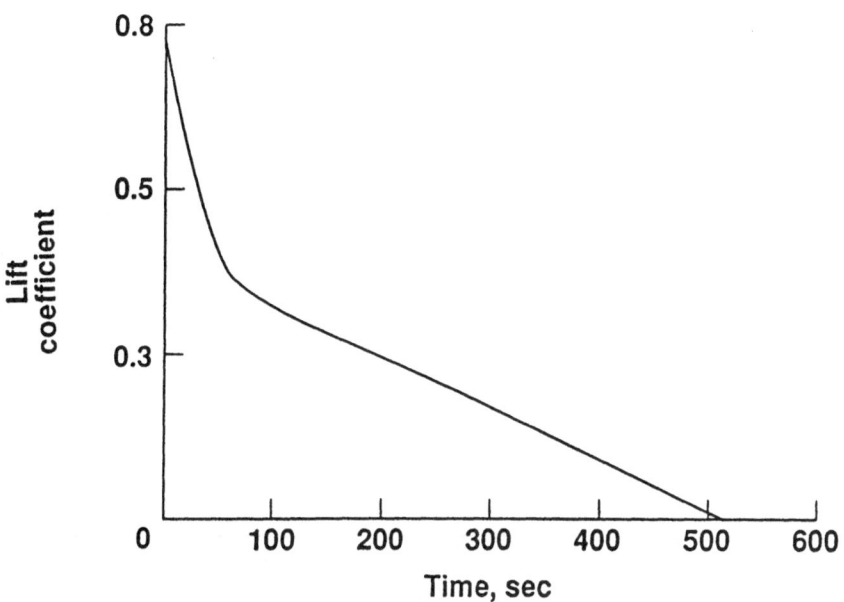

Fig. 6(a)   Time History of Lift Coefficient

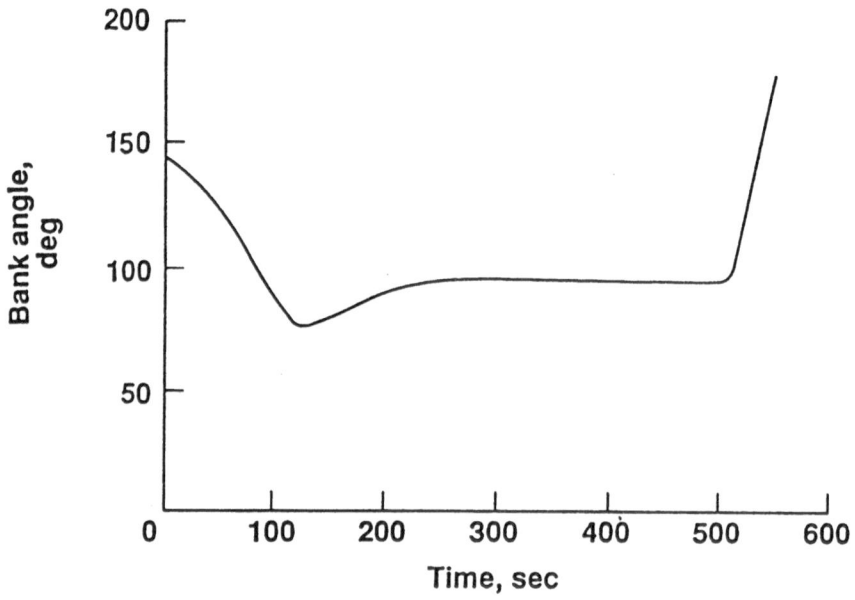

Fig. 6(b)   Time History of Bank Angle

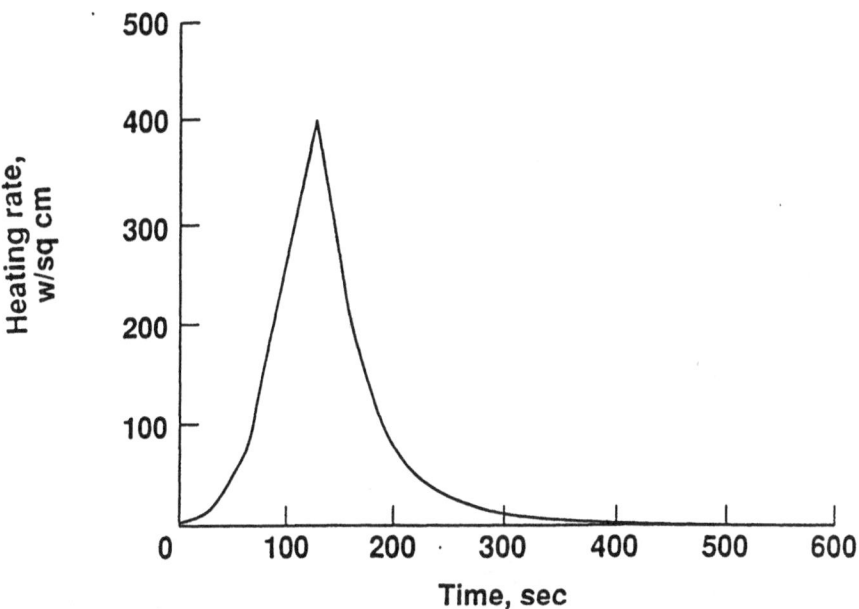

Fig. 7(a) Time History of Heating Rate

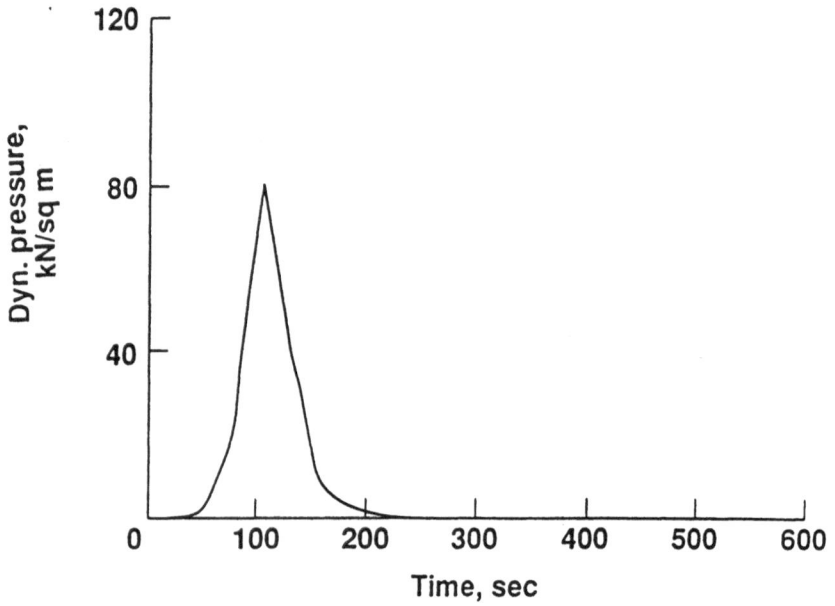

Fig. 7(b) Time History of Dynamic Pressure

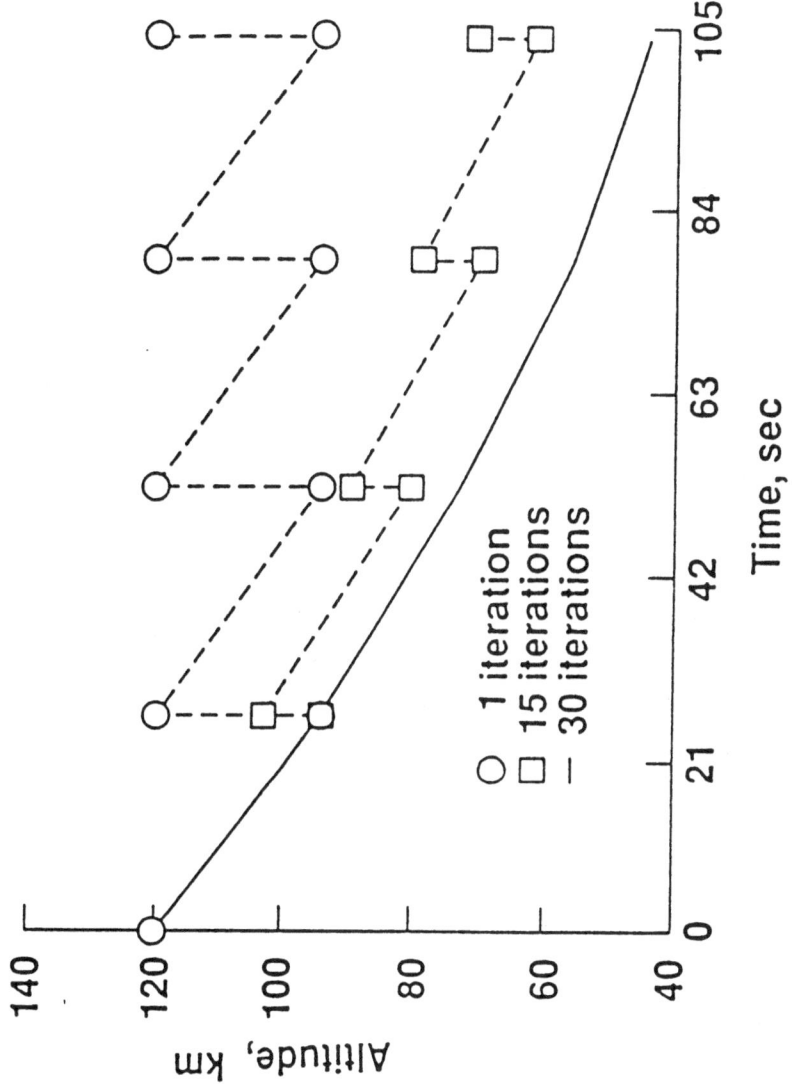

Fig. 8   Successive Approximations for Altitude

4.5 Concluding Remarks

This Chapter has addressed the problem of minimization of fuel consumption during the atmospheric portion of an aeroassisted, orbital transfer with plane change. The complete mission has required three characteristic velocities, a deorbit impulse at high Earth orbit, a boost impulse at the atmospheric exit, and a reorbit impulse at low Earth orbit. A performance index has been formulated as the sum of these three impulses. Application of optimal control principles has led to a nonlinear, two-point, boundary value problem which was solved by using a multiple shooting algorithm. For a typical data, the simulations have been obtained.

Nomenclature

$A_1 = C_{DO} S \rho_s H_a / 2m$

$A_2 = C_{LR} S \rho_s H_a / 2m$

AOTV : aeroassisted orbital transfer vehicle

$b = R_a / H_a$

C_D: drag coefficient

C_{DO}: zero-lift drag coefficient

C_L : lift coefficient

C_{LR}: lift coefficient for maximum lift-to-drag ratio

D :   drag force

E_m : maximum value of L/D

g :   gravitational acceleration

GEO : geosynchronous Earth orbit

H :   altitude

HEO : high Earth orbit

$\mathcal{H}$: Hamiltonian

i : inclination

J : performance index

K : induced drag factor

L : lift force

LEO : low Earth orbit

m : vehicle mass

OTV : orbital transfer vehicle

R : distance from Earth center to vehicle center of gravity

R_a: radius of atmospheric boundary

R_c : radius of low Earth orbit

R_d : radius of high Earth orbit

R_E : radius of Earth

S : aerodynamic reference area

SSO : space station orbit

t : time

TPBVP : two-point boundary value problem

V : velocity

v : normalized velocity

β : inverse atmospheric scale height

γ : flight path angle

ψ : heading angle

σ : bank angle

θ : down range angle

φ : cross range angle

δ : normalized density

λ : costate (Lagrange) variable

μ : gravitational constant of Earth

$\eta = C_L/C_{LR}$

ρ : density

τ : normalized time

ΔV : characteristic velocity

Δv : normalized characteristic velocity

Subscripts

c : circularization or reorbit at LEO

d : deorbit at HEO

e : entry to atmosphere

f : exit from atmosphere

s : surface level

References

[1] G. D. Walberg, "A Survey of Aeroassisted Orbital Transfer," *Journal of Spacecraft*, 22, 3-18, 1985

[2] N. X. Vinh, *Optimal Trajectories in Atmospheric Flight.* Amsterdam: Elsevier Scientific Publishing Co., 1981.

[3] E. D. Dickmann, *The effect of finite thrust and heating constraints on the synergetic plane change maneuver for space-shuttle orbiter-class vehicle*, NASA TN D-7211, Oct. 1973.

[4] D. G. Hull, J. M. Glitner, J. L. Speyer, J. and Maper, "Minimum Energy Loss Guidance for Aeroassisted Orbital Plane Change," *Journal of Guidance, Control, and Dynamics*, 8, 487-493, 1985.

[5] N. X. Vinh, and J. M. Hanson, "Optimal Aeroassisted Return from High Earth Orbit with Plane Change," *Acta Astronautica*, 12, 11-25, 1985.

[6] A. Miele, V. K. Baspur, and W. Y. Lee, "Optimal Trajectories for Aeroassisted Noncoplanar Orbital Transfer," *Acta Astronautica*, 15, 399-412, 1987.

[7] D. S. Naidu, J. L. Hibey, and C. Charalambous, "Fuel-optimal trajectories for aeroassisted coplanar orbital transfer problem," *IEEE Transactions on Aerospace and Electronic Systems*, 26, 374-381, 1990.

[8] H. B. Keller, *Numerical Methods for Two-Point Boundary-Value Problems*, Waltham, MA: Blaisdell Publ. Co., 1968.

[9] J. Stoer, and R. Bulirsch, *Introduction to Numerical Analysis*, New York: Springer-Verlag, 1980.

[10] H. J. Pesch, "Numerical computation of neighboring optimum feedback control schemes in real-time," *Applied Mathematics and Optimization*, 5, 231-252, 1979.

[11] J. P. Marec, *Optimal Space Trajectories*, Amsterdam: Elsevier Scientific Publishing Company, 1979.

[12] T. D. Talay, N. H. White, and J. C. Naftel, "Impact of Atmospheric Uncertainties and Viscous Interaction Effects on the Performance of Aeroassisted Orbital Transfer Vehicles," *AIAA 22nd Aerospace Science Meeting*, Reno, NV, Jan. 1984.

****

CHAPTER 5

ORBITAL PLANE CHANGE WITH AEROCRUISE

5.1 Introduction

It has been established that a significant fuel savings and hence increased payload capabilities can be achieved with propulsive and aeroassisted maneuvers instead of all-propulsive maneuvers [1]. This leads to an aeroassisted orbital transfer vehicle (AOTV), which on its return leg of the mission, dips into the Earth's atmosphere, utilizes atmospheric drag to reduce orbital velocity and to achieve a desired orbital inclination. Basically, the AOTV performs a synergistic maneuver, employing a hybrid combination of propulsive maneuver in space and aerodynamic maneuver in the atmosphere [1].

The plane change capability is required to (i) orbit a vehicle in a plane which does not pass through a launch site, (ii) shorten the time needed to reach multiple reconnaissance targets on a single orbital mission, (iii) reduce the time needed to return to base from orbit, (iv) perform effective rendezvous with satellites in different orbital planes, (v) avoid flights over hostile territory, and finally (vi) facilitate arrival and departure flights from Space Station Freedom, in fulfilling specified mission objectives [2]. It should be noted that an orbital plane is usually defined in terms of inclination and longitude of the ascending node. For the present purpose only an inclination change is controlled.

There are basically three methods of plane change, (i) impulsive method, (ii) aeroglide method, and (iii) aerocruise method. In impulsive method, the plane change is achieved entirely outside the atmosphere, and fuel consumption is prohibitively large for sizable changes of orbital plane. In both aeroglide and aerocruise methods, rockets are used to deflect the vehicle into the atmosphere, and the plane change is accomplished by heading change of the vehicle. With aeroglide there is no thrusting during the atmosphere, and with aerocruise, atmospheric drag is balanced by a continuous thrust to keep the spacecraft at a constant altitude and

This Chapter is based on D. S. Naidu, "Orbital plane change with aerocruise," Originally presented at AIAA 29th Aerospace Sciences and Meeting and Exhibit, Reno, Nevada, Jan. 7-10, 1991. Permission given by AIAA is hereby acknowledges.

velocity. Propellant expenditure comparisons among the three methods of plane change show that the aerocruise method is superior to other competing methods for plane changes greater than about 20 degrees, and with heating restraints. The basic effect of propulsion during aerocruise is to (i) balance drag in order to maintain constant velocity, (ii) augment lift with a component of thrust, thus increasing cruising altitude over what it would be during aeroglide turn, and finally (iii) provide a lateral component of thrust giving the required turn necessary for plane change. The aeroglide and aerocruise methods utilizing atmospheric maneuver in conjunction with propulsion augmentation are also termed the "synergistic" or "aeropropulsive" methods [3,4].

The following are some of the features of atmospheric plane change [5-7]. (i) For plane changes of less than 15 degrees, an all-propulsive maneuver is generally more efficient. (ii) An L/D of at least 2 is required to offer a significant advantage over the all-propulsive plane change, and it is desirable to maximize the L/D of a vehicle. (iii) A plane change made at a node produces all inclination change whereas a turn at an orbit apex (90 degrees from node) provides no inclination change, only a shift in the node. Hence, for maximum inclination change and minimum node shift, the turn should be centered over the node in the shortest possible duration. Thus, plane changes performed at maximum C_L (i.e., high angle of attack) which are quicker are more fuel efficient than the slower maximum L/D turns. (iv) The total heat load can be reduced substantially by carrying out a quicker high angle-of-attack turns rather than the slower maximum L/D turns. (v) An aerocruise (thrusting) turn offers significant advantages over an aeroglide (non-thrusting) turn [9,10], when the desired plane change is more than 10 degrees. (vi) During aerocruise, the high angle-of-attack and bank attitude of the vehicle produce a lateral component of thrust, which is responsible for a significant amount of plane change.

In this Chapter, consider the synergistic plane change problem arising in noncoplanar orbital transfer employing aeroassist technology. The mission involves the transfer from HEO to LEO with plane change being performed within the atmosphere. The complete mission consists of a deorbit phase, an atmospheric phase, and finally a reorbit phase. The atmospheric maneuver phase is composed of descent (entry) mode, cruise mode, and ascent (exit) mode. During the aerocruise mode, constant altitude and velocity are maintained either by (i) varying bank angle with constant thrust, or by (ii) varying thrust with constant bank angle. The comparison of these two control schemes bring out some interesting features. Numerical results are given for typical data.

5.2 Mission Description

For an orbital transfer problem, the following assumptions are made. (i) The initial HEO and final LEO orbits are circular. (ii) The mission is comprised of three impulses. (iii) The vehicle is represented as a constant point mass during atmospheric pass. (iv) A Newtonian inverse square gravitational field is used. (v) Earth's rotation is neglected. (vi) The atmosphere is exponential.

The complete mission from HEO to LEO with atmospheric pass is depicted in Fig. 1. It consists of a deorbit phase, an atmospheric phase, and a reorbit phase. There are three impulses: first, a deorbit impulse ΔV_d at HEO to inject a vehicle into a HEO-entry elliptic orbit, second, a boost impulse ΔV_b at the exit from the atmosphere for the vehicle to attain sufficient velocity to travel along an exit-LEO elliptic orbit, and finally, a circularizing impulse ΔV_c to circularize the path of the vehicle. The atmospheric phase itself is composed of descent (entry) mode, cruise mode, and ascent (exit) mode.

Consider the basic equations of motion for different phases of deorbit, aeroassist (or atmospheric flight), boost and reorbit (or circularization).

5.3 Deorbit Phase

Initially, assume that a spacecraft is in a circular orbit of radius R_d, well outside the Earth's atmosphere, moving with a circular velocity $V_d = \sqrt{\mu/R_d}$. Deorbit is performed by means of an impulse ΔV_d, to transfer the vehicle from the circular orbit to elliptic orbit with perigee low enough to intersect the dense part of the atmosphere [Fig. 1]. At D, since the elliptic velocity is less than the circular velocity, the impulse ΔV_d is executed so as to oppose the circular velocity V_d. The deorbit impulse ΔV_d causes the vehicle to enter the atmosphere of radius R_a with a velocity V_e and flight path angle γ_e. It is known that the optimal-energy loss maneuver from the circular orbit is simply the Hohmann transfer and the impulse is parallel and opposite to the instantaneous velocity vector.

Using the principle of conservation of energy and angular momentum at the deorbit point D, and the atmospheric entry point E yields [11],

$$V_e^2/2 - \mu/R_a = (V_d - \Delta V_d)^2/2 - \mu/R_d$$

$$R_a V_e \cos(-\gamma_e) = R_d(V_d - \Delta V_d)$$

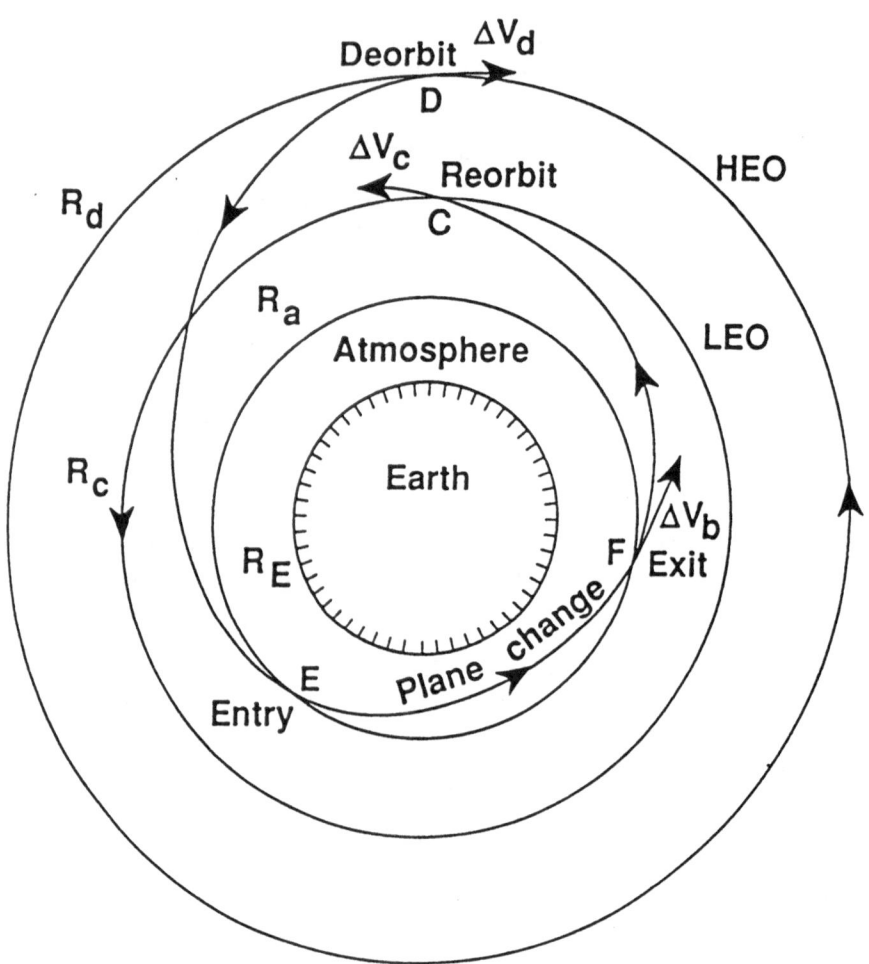

Fig. 1   Aeroassisted Orbital Transfer Mission

$$\Delta V_d = \sqrt{\mu/R_d} - \sqrt{2\mu(1/R_a - 1/R_d)/[(R_d/R_a)^2/\cos^2\gamma_e - 1]}$$

It is easily seen that the minimum value of the deorbit impulse ΔV_{dm} obtained at $\gamma_e = 0$, corresponds to an ideal transfer wherein the space vehicle grazes along the atmospheric boundary. To ensure proper atmospheric entry, the deorbit impulse ΔV_d must be higher than the minimum deorbit impulse ΔV_{dm} which is given by

$$\Delta V_{dm} = \sqrt{\mu/R_d} - \sqrt{2\mu(1/R_a - 1/R_d)/[(R_d/R_a)^2 - 1]}$$

5.4 Aeroassist Phase

The atmospheric phase of the mission is composed of (i) descent mode, (ii) cruise mode, and (iii) ascent mode [Fig. 2].

5.4.1 Descent Mode

During the descent mode, the equations of motion for the vehicle (without any thrusting) are given as [Fig. 3], [12],

$$\frac{dR}{dt} = V\sin\gamma$$

$$\frac{d\theta}{dt} = V\cos\gamma\cos\psi/R\cos\phi$$

$$\frac{d\phi}{dt} = V\cos\gamma\sin\psi/R$$

$$m\frac{dV}{dt} = -D - mg\sin\gamma$$

$$m\frac{dV}{dt} = L\cos\sigma + m(V/R - g)\cos\gamma$$

$$mV\frac{d\psi}{dt} = L\sin\sigma/\cos\gamma - (mV^2/R)\cos\gamma\cos\psi\tan\phi \qquad (1)$$

where,

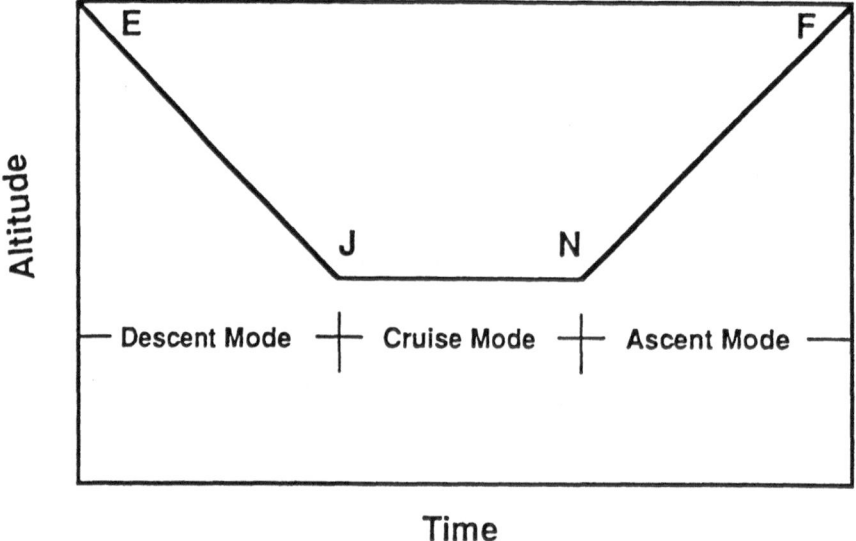

Fig. 2   Atmospheric Phase

$$L = C_L^2(\alpha)\rho S V \ /2; \quad D = C_D^2(\alpha)\rho S V \ /2; \quad C_D = C_{DO}^2 + KC_L$$

$$g = \mu/R_2; \quad R = H + R_E; \quad \rho = \rho_s \exp(-\beta H)$$

Neglecting mass terms in comparison to aerodynamic terms in (1) gives

$$\frac{dH}{dt} = V\sin\gamma$$

$$\frac{dV}{dt} = - D/m$$

$$\frac{d\gamma}{dt} = L\cos\sigma/mV$$

$$\frac{d\psi}{dt} = L\sin\sigma/mV\cos\gamma \tag{2}$$

Using flight path angle as the independent variable,

$$\frac{dH}{d\gamma} = 2m\exp(\beta H)\sin\gamma/\rho_0 S C_L \cos\sigma$$

$$\frac{dV}{d\gamma} = - C_D V/C_L \cos\gamma$$

$$\frac{d\psi}{d\gamma} = \tan\sigma/\cos\gamma \tag{3}$$

During the descent mode, let us assume that there is no banking and hence no heading. Then (3) becomes

$$\frac{dH}{d\gamma} = 2m\exp(\beta H)\sin\gamma/\rho_0 S C_L$$

$$\frac{dV}{d\gamma} = - C_D V/C_L \tag{4}$$

Optimal Control

The optimal control problem is posed as follows. Given entry conditions, and the conditions at the end of the descent mode (or the initiation of aerocruise mode), find the optimal control law which maximizes the final velocity, subject to altitude constraint $H \geq H_b$[13], This altitude constraint implies in a way heat-rate constraint. The performance index is

$$J = -V_b$$

The Hamiltonian for the above system becomes

$$\mathcal{H} = \lambda_H \left[2m\exp(\beta H)\sin\gamma/\rho_0 SC_L \right] + \lambda_v \left[-C_D V/C_L \right]$$

The adjoint equations are

$$d\lambda_H/d\gamma = -\lambda_H 2m\beta\exp(\beta H)\sin\gamma/\rho_0 SC_L$$

$$d\lambda_v/d\gamma = \lambda_v C_D/C_L$$

Solving the state and costate equations,

$$V\lambda_v = V_e \lambda_{ve} = \text{constant}$$

$$\lambda_H \exp(\beta H) = \lambda_{He} \exp(\beta H_e)$$

The boundary conditions for adjoint variables are

$$\lambda_v(\gamma = \gamma_j) = \left.\frac{\partial J}{\partial V}\right|_{\gamma = \gamma_j} = -1$$

$$\lambda_H(\gamma = \gamma_j) = \left.\frac{\partial J}{\partial H}\right|_{\gamma = \gamma_j} = 0$$

From the above,

$$\lambda_H = 0$$

With this, the Hamiltonian reduces to

$$\mathcal{H} = -\lambda_v V[C_{D0}/C_L + KC_L]$$

The optimal control is then given by

$$\frac{\partial \mathcal{H}}{\partial C_L} = 0$$

leading to

$$C_{LO} = \sqrt{C_{DO}/K} = C_{LE}$$

where, C_{LE} is the lift coefficient for maximum lift-to-drag ratio $(L/D)_{max} =$ $E = 1/2\sqrt{KC_{DO}}$. With the above optimal control in (4), solve for the velocity and altitude as

$$V(\gamma) = V_e \exp[-(\gamma - \gamma_e)/E]$$

$$H = \ln\left[\exp(-\beta H_e) + \frac{2\beta m}{\rho_0 SC_{LO}}(\cos\gamma - \cos\gamma_e)\right]^{-1/\beta}$$

For small γ, this reduces to

$$H = \ln\left[\exp(-\beta H_e) - \frac{\beta m}{\rho_0 SC_{LO}}(\gamma^2 - \gamma_e^2)\right]^{-1/\beta}$$

At the start of the descent mode, $\gamma = \gamma_e$, and at the end of the descent mode, $\gamma = \gamma_j = 0$. Then the above relations become

$$H_j = \ln\left[\exp(-\beta H_e) + \frac{2\beta m}{\rho_0 SC_{LO}}(1 - \cos\gamma_e)\right]^{-1/\beta}$$

and with the approximation

$$H_j = \ln\left[\exp(-\beta H_e) + \frac{\beta m}{\rho_0 SC_{LO}}\gamma_e^2\right]^{-1/\beta}$$

Then, the inequality constraint on altitude $H \geq H_j$ transforms to

$$\gamma_e \leq \cos^{-1}\left[1 - \frac{\rho_0 S C_{L0}}{2m\beta}\left\{\exp(-\beta H_j) - \exp(-\beta H_e)\right\}\right]$$

and with approximate solution,

$$\gamma_e^2 \leq \frac{\rho_0 S C_{L0}}{m\beta}\left[\exp(-\beta H_j) - \exp(-\beta H_e)\right] \tag{5}$$

The velocity at the end of the descent mode is obtained with $\gamma = \gamma_j = 0$ as

$$V_j = V_e \exp\left(2\gamma_e\sqrt{C_{D0}K}\right) \tag{6}$$

5.4.2 Aerocruise Mode: Bank Angle Control

First write down the general equations of motion, inject the conditions for cruise flight, use the assumptions of small latitude, and finally optimize the heading change. During aerocruise mode, there is continuous thrusting [12]. Thus the kinematic equations are [Fig. 3]

$$\frac{dH}{dt} = V\sin\gamma$$

$$\frac{d\theta}{dt} = V\cos\gamma\cos\psi/R\cos\phi$$

$$\frac{d\phi}{dt} = V\cos\gamma\sin\psi/R$$

The force equations are

$$m\frac{dV}{dt} = T\cos\eta - D - mg\sin\gamma$$

$$mV\frac{d\gamma}{dt} = (T\sin\eta + L)\cos\sigma + m(V^2/R - g)\cos\gamma$$

$$mV\frac{d\psi}{dt} = (T\sin\eta + L)\sin\sigma/\cos\gamma - (mV^2/R)\cos\gamma\cos\psi\tan\phi$$

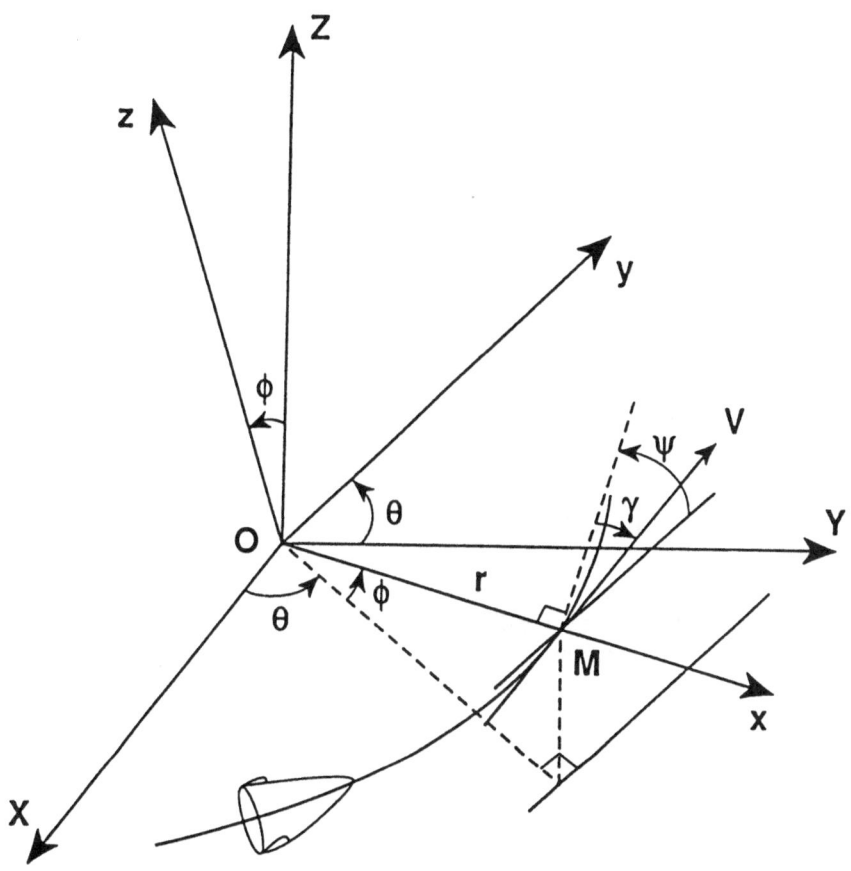

Fig. 3 Coordinate System

The propulsion (thrusting) equation is

$$\frac{dm}{dt} = -T/gI_{sp} \tag{7}$$

From the above equations of motion, we see clearly that during the atmospheric maneuver, if the lift vector L is rotated about the velocity vector V through the bank angle σ, it creates a lateral force component $(T\sin\eta + L)\sin\sigma$ orthogonal to the vertical plane that has the effect of changing the heading angle ψ. At the end of the atmospheric phase, the relations for the cross range angle ϕ, and the heading angle ψ, become [12],

$$\frac{d\phi/dt}{d\psi/dt} = -\frac{\tan\psi}{\tan\phi}$$

integration of which yields,

$$\cos\phi\cos\psi = \cos i$$

where, i is the orbital inclination. For small values of cross range angle ϕ, the orbital inclination i is given by the heading angle ψ itself. Thus, the total change in the heading corresponds to the change in orbital inclination (plane change).

Now insert the cruise conditions of constant altitude and velocity [8]. The constant altitude condition gives zero flight path angle throughout. The constant velocity condition on boils down to

$$T\cos\eta = D = \rho S V^2 C_D(\alpha)/2$$

Note that the conditions at the beginning of the aerocruise mode are denoted by the subscript j. However, for simplicity in notation we shall continue to use the variables without any subscript to denote the cruise conditions. If the angle of attack α is held constant, then the drag force D is constant at a constant cruising altitude. Also, since the flight path angle is zero throughout, (7) reduces to

$$(T\sin\eta + L)\cos\sigma = m(g - V^2/R) \tag{8}$$

Combining the above two equations, we get

$$\cos\sigma = \frac{m(g - V^2/R)\cos\eta}{D[\sin\eta + (L/D)\cos\eta]} \tag{9}$$

where,

$$\sin\eta + (L/D)\cos\eta = (L + T\sin\eta)/T = E_p$$

is called aeropropulsive efficiency. Also,

$$(D\tan\eta + L)\cos\sigma = m(g - V^2/R)$$

The above equation is also rewritten as

$$\tan\eta = mK_1/\cos\sigma - L/D \tag{10}$$

where, $K_1 = (g - V^2/R)/D = (gR - V^2)/RD$

From (8), we see that for a given angle of attack α, if altitude H, and velocity V are kept constant, then the drag D and lift L forces are constant. The mass m always changes due to thrusting. Then the above relation (10) for variable mass can be satisfied in any one of the following three ways.

(i) Variable bank angle with constant angle of attack and thrust angle: With bank control, (7) and (10) mean that the thrust T is constant leading to a constant mass flow rate.

(ii) Variable thrust with constant bank angle: On the other hand, with thrust control, (8) implies that we need to change both magnitude and angle of the thrust, in order to keep a constant drag force. Thus for cruise condition, both thrust magnitude and angle need to be controlled such that $T\cos\eta$ is constant, but $T\sin\eta$ changes according to (8) [see Fig. 3]. This leads to variable mass flow rate.

(iii) Variable bank angle and variable thrust: Here, we change both bank angle and thrust magnitude and angle, in order to satisfy the cruise conditions (9). This also leads to variable mass flow rate.

Obviously, the bank angle control leading to constant thrust (and hence constant mass flow rate) seems to be the simplest of all for implementation. However, it will be interesting to see which of the control schemes provides greatest amount of heading change and thereby inclination for the same amount of fuel expenditure.

The bank angle control with constant thrust magnitude and angle has been thoroughly discussed using arc length as independent variable [8]. However, in our present work, we continue to use time as independent variable. Assuming the latitude to be small, the cruise motion is described by

$$\frac{d\theta}{dt} = V\cos\psi/R$$

$$\frac{dm}{dt} = - K_2/\cos\eta$$

$$\frac{d\psi}{dt} = K_3\tan\sigma \tag{11}$$

$$K_2 = D/gI_{sp} \; ; \quad K_3 = g/V - V/R$$

The bank angle control is given by

$$\cos\sigma = mK_1/[\tan\eta + L/D] = mK_4 \tag{12}$$

$$K_4 = \frac{(g - V^2/R)}{(L + D\tan\eta)} = \frac{2(k^2 - 1)}{\rho RSC_D(\alpha)[\tan\eta + L/D]}$$

and $k = \sqrt{gR}/V$, the ratio of circular speed to cruise speed at R. From (12), we see that for a given angle of attack, and at constant altitude, speed, and thrust angle, the bank angle has to be varied as per the mass. That is, as mass m decreases along the flight, bank angle σ should be increased. Thus, in increasing the bank angle with the decrease of mass, we are trying to balance the decrease in the difference between the vehicle's weight and centrifugal force with the sum of vertical components of lift and thrust. In this control scheme, both mass and bank angle change, whereas altitude, velocity, angle of attack, thrust, thrust angle, and mass flow rate are held constant.

The cruise condition (12) reveals that (a) With $\sigma = 0^0$, there is no banking and the cruise conditions can be maintained only by variable thrusting. (b) For $0 < \sigma < 90^0$, the cruising speed is less than circular speed. The lift is directed upward. The gravitational force is higher than the centrifugal force. (c) $\sigma = 90^0$ corresponds to cruise speed being equal to the circular speed, and all the aerodynamic force ($T\sin\eta + L$) is used for heading change or turning. The gravitational force is equal to the centrifugal force.(d) For $\sigma > 90^0$, the cruising speed is higher than the circular speed. The centrifugal force is higher than the gravitational force and hence the lift is directed downward in order to prevent the vehicle to escape from Earth.

Given the initial values of mass m_j, and heading angle ψ_j, find the initial bank angle σ_j from (29a). Also, (28b) can be solved directly as

$$m(t) = - (K_2/\cos\eta)t + m_j \tag{13}$$

Thus, $\sigma(t)$, and $\psi(t)$ are solved until either of the desired final conditions m_n, or ψ_n is realized. With a constant thrust angle η, and given initial mass m_j, and heading angle ψ_j, and final mass m_n (or final heading angle ψ_n) the sequence of solution of the aerocruise problem is to solve, first the mass equation (11), second the bank angle equation (12), and finally the heading angle equation (11).

In this formulation for aerocruise, see that the heading angle changes with respect to bank angle as given by (11), and bank angle in turn has to follow the mass as per (12), and the mass varies independently according to (13). Hence, there is no optimization of heading angle w.r.t. bank angle control variable, for a given fuel consumption or of fuel consumption w.r.t. bank angle control variable, for a given heading change. Alternatively, from (12) and (13), solve for ψ, which is now a function of thrust angle η, cruising altitude H, and cruising speed V. Then, find the optimal value of η, which should be maintained constant throughout the aerocruise to achieve maximum ψ.

Optimization of Heading Change w.r.t. Thrust Angle

The optimization problem here is to find an optimum thrust angle which is to kept constant throughout the cruise mode, in order to maximize the heading change. We can solve this problem in a variety of ways. Basically, the heading angle ψ can be solved from (11) to (13) in terms of time t, mass m, or bank angle σ. Thus,

$$\psi_n = \psi_j + K_5\left[\ln(A_n/A_j) + B_j - B_n\right] \tag{14}$$

where,

$$K_5 = \frac{I_{sp}g}{V}\left[\sin\eta + (L/D)\cos\eta\right]$$

$$A_i = (1 + \sin\sigma_i)/\cos\sigma_i; \quad B_i = \sin\sigma_i; \quad i = j, n$$

Note that A_i, and B_i can also be expressed in terms of mass m using (12), or in terms of time t with (13). For example, in terms of mass m, (14) becomes

$$A_i = \frac{\left[1 + \sqrt{1 - (m_i K_4)^2}\right]}{m_i K_4} \; ; \quad B_i = \sqrt{1 - (m_i K_4)^2}$$

In terms of bank angle σ, (14) is rewritten as [13],

$$\psi_n = \psi_j + \frac{I_{sp} g}{V}\left[\sin\eta + (L/D)\cos\eta\right]\left[\ln\left\{\frac{\cos\sigma_j(1 + \sin\sigma_n)}{\cos\sigma_n(1 + \sin\sigma_j)}\right\} + \sin\sigma_j - \sin\sigma_n\right]$$

(15)

where, bank angle σ is related with thrust angle η as per (12). Now optimizing ψ_n with respect to thrust angle η, (i.e., making $d\psi_n/d\eta = 0$) and assuming the initial value $\psi_j = 0$ yields

$$\cos\eta[\cos\eta - (L/D)\sin\eta]\left[\ln\left\{\frac{\cos\sigma_j(1 + \sin\sigma_n)}{\cos\sigma_n(1 + \sin\sigma_j)}\right\} + \sin\sigma_j - \sin\sigma_n\right] = 0$$

Using first order approximations in the change of the bank angle,

$$\cos\sigma_n = \cos(\sigma_j + \Delta\sigma) \cong \cos\sigma_j - (\sin\sigma_j)\Delta\sigma$$

$$\sin\sigma_n = \sin(\sigma_j + \Delta\sigma) \cong \sin\sigma_j + (\cos\sigma_j)\Delta\sigma$$

and, linearizing the logarithm, the above transcendental equation becomes,

$$\left[\cos\eta[\cos\eta + (L/D)\sin\eta]\sin^2\sigma_j + \cos^2\sigma_j\right](1 + \sin\sigma_j)\Delta\sigma = 0$$

Assuming that $1 + \sin\sigma_j \neq 0$, and $\Delta\sigma \neq 0$, the above equation becomes

$$\cos\eta[\cos\eta + (L/D)\sin\eta]\sin^2\sigma_j + \cos^2\sigma_j = 0$$

(16)

From (12),

$$\cos\sigma_j = \frac{A}{\tan\eta + L/D}; \quad A = m_j K_1$$

Using the above in (16),

$$[\tan\eta + L/D]\left[(L/D)\tan^2\eta + \{(L/D)^2 - A^2 - 1\}\tan\eta - L/D\right] = 0$$

Again, assuming $\tan\eta + L/D \neq 0$, we finally get a simplified form as [13],

$$(L/D)\tan^2\eta + [(L/D)^2 - A^2 - 1]\tan\eta - L/D = 0 \tag{17}$$

The implication of $\tan\eta + L/D = 0$ is that (a) $90^0 < \eta < 180^0$, or (b) $270^0 < \eta < 360^0$. The first condition implies that the thrust T will aid the drag force instead of opposing it, and the second condition shows that the thrust T will oppose the lift instead of aiding it. Summarizing, for bank angle control we have

$$\frac{d\psi}{dt} = K_3 \tan\sigma$$

$$\frac{dm}{dt} = -K_2/\cos\eta$$

The cruise condition is given by

$$\cos\sigma = mK_1/[\tan\eta + L/D] \tag{18}$$

The sequence of solutions is (i) Using initial masses m_j, and L/D, solve (17) for optimal thrust angle η. (ii) Using η, m_j, and L/D, solve (18) for σ_j. (iii) Using η, and m_j, solve for new m from (18). (iv) Using the mass m, and η, solve (18) or (14) for new ψ. (v) Go to (ii), and repeat the steps. (vi) Integration stops when m reaches the final m_n in (18) and the corresponding maximum heading angle ψ_n is obtained from (18).

Although (18) can be solved for either set of given values of (a) m_j and m_n or (b) ψ_j and ψ_n, the condition (17) requires m_j and m_n to determine the optimal η which is kept constant throughout the aerocruise mode. Hence, given the fuel consumption $(m_j - m_n)$, try to determine the maximum heading angle change.

5.4.3 Aerocruise Mode: Thrust Control

Here, keep the bank angle constant throughout and change the thrust magnitude and angle in order to achieve the desired heading change and hence the inclination [14]. For the sake of simplicity, repeat the equations at cruise,

$$\frac{d\theta}{dt} = V\cos\psi/R$$

$$\frac{dm}{dt} = -K_2/\cos\eta$$

$$\frac{d\psi}{dt} = K_3\tan\sigma \tag{19}$$

The cruise conditions are given by

$$T\cos\eta = D$$

$$(T\sin\eta + L)\cos\sigma = m(g - V^2/R) \tag{20}$$

Combining the two conditions,

$$\tan\eta = mK_1/\cos\sigma - L/D \tag{21}$$

For a constant altitude H, speed V, and given angle of attack α and bank angle σ, as the mass m changes, the thrust angle η follows (21). At the same time, the drag force D is to be kept constant as per (20). Thus, in order to satisfy both the conditions (20), adjust thrust magnitude T and angle η in such a way that $T\cos\eta$ is kept constant, and $T\sin\eta$ changes as per mass m. From (19), see that the bank angle is kept constant throughout the cruise mode and hence the rate of change of heading angle is constant, whereas the mass flow rate is variable. This is in contrast to the bank angle control discussed in the last section.

Given a constant bank angle σ, the initial and final conditions m_j, ψ_j, m_n (or ψ_n), the sequence of solutions for the cruise flight with thrust control is first, solve the cruise condition equation (21) for η, second solve mass rate equation (19) for m, and finally solve the heading angle equation (19) for ψ. In the next section try to find the optimal bank angle which should be kept constant throughout the cruise, to get maximum heading change.

Optimization of Heading Angle w.r.t. Bank Angle

Here, the interest is in finding the optimum bank angle so that the heading change is maximized. For this, first solve (19) to (21) for the heading angle and then find the stationary value of ψ w.r.t. σ. Thus,

$$\psi_n = (I_{sp}g/V)\sin\sigma\ln\left[\frac{\sec\eta_j + \tan\eta_j}{\sec\eta_n + \tan\eta_n}\right] \tag{22}$$

where,

$$\tan\eta_i = m_iK_1/\cos\sigma - L/D; \quad \sec^2\eta_i = 1 + \tan^2\eta_i$$

$$K_6 = (I_{sp}g/V)\sin\sigma$$

Alternatively, (22) can be used to find the mass m for a given ψ. Use of the stationary condition leads to

$$\cos^2\sigma\ln\left[\frac{\cos\eta_j(1 + \sin\eta_n)}{\cos\eta_n(1 + \sin\eta_j)}\right] + \sin^2\sigma\left[(\sin\eta_n - \sin\eta_j) + \right.$$

$$\left. (L/D)(\cos\eta_n - \cos\eta_j)\right] = 0$$

Considering only the first order approximations in η

$$\cos\eta_n = \cos(\eta_j + \Delta\eta) \cong \cos\eta_j - (\sin\eta_j)\Delta\eta$$

$$\sin\eta_n = \sin(\eta_j + \Delta\eta) \cong \sin\eta_j + (\cos\eta_j)\Delta\eta$$

and linearizing the logarithm, the above stationary condition becomes

$$\left\{\cos^2\sigma + \sin^2\sigma\left[\frac{1 - (L/D)\tan\eta_j}{1 + \tan^2\eta_j}\right]\right\}[1 + \sin\eta_j]\Delta\eta = 0$$

Assuming $1 + \sin\eta_j \neq 0$; and $\Delta\eta \neq 0$, we have

$$\cos^2\sigma + \sin^2\sigma \left[\frac{1 - (L/D)\tan\eta_j}{1 + \tan^2\eta_j} \right] = 0$$

Using

$$\tan\eta_j = A/\cos\sigma - L/D$$

a simplified quadratic equation in $\cos\sigma$ is obtained as

$$(L/D)A\cos^2\sigma - [(L/D)^2 + A^2 + 1]\cos\sigma + (L/D)A = 0 \tag{23}$$

It is interesting to note that the optimal conditions (17) and (23) for bank control and thrust control respectively, are interchangeable by the cruise condition.

Summarizing, for thrust control we have

$$\frac{d\psi}{dt} = K_3\tan\sigma$$

$$\frac{dm}{dt} = -K_2/\cos\eta \tag{24}$$

The cruise conditions are given by

$$T\cos\eta = D$$

$$\tan\eta = mK_1/\cos\sigma - L/D \tag{25}$$

The sequence of solutions is (i) Using initial mass m_j and L/D, solve (23) for optimal bank angle σ. (ii) Using σ, m_j, and L/D, solve (25) for η_j. (iii) Using η_j, find the thrust T_j from (25). (iv) Using ψ_j, and σ, solve (50) for new ψ. (v) Using η_j, and m_j, solve for new m from (24) or (22). (vi) Go to (ii), and repeat the steps. (vii) Integration stops when m reaches the final m_n in (24) and the corresponding maximum heading angle ψ_n

is obtained from (24).

Although (24) and (25) can be solved for either set of given values of (a) m_j and m_n or (b) ψ_j and ψ_n, the condition (23) requires m_j and m_n to determine the optimal σ which is kept constant throughout the aerocruise mode. Hence, given the fuel consumption $(m_j - m_n)$, try to determine the maximum heading angle change.

5.4.4 Ascent Mode

This is just a replica of the descent mode except for the change in the mass of the vehicle, and the boundary conditions. Thus, consider the equations of motion, change the independent variable to flight path angle, and finally assume that there is no appreciable change in heading angle. Thus,

$$\frac{dH}{d\gamma} = 2m_n \exp(\beta H)\sin\gamma/\rho_0 SC_L$$

$$\frac{dV}{d\gamma} = -C_D V/C_L \qquad (26)$$

The optimal control problem is posed as follows. Given initial conditions (or the conditions at the end of the cruise mode), and the conditions at the end of the ascent mode, find the optimal control law which maximizes the final velocity [12]. The performance index is given by

$$J = -V_f$$

The Hamiltonian for the above becomes

$$\mathcal{H} = \lambda_H \left[2m_n \exp(\beta H)\sin\gamma/\rho_0 SC_L \right] + \lambda_v \left[-C_D V/C_L \right]$$

The adjoint equations are

$$d\lambda_H/d\gamma = -\lambda_H 2m_n \beta \exp(\beta H)\sin\gamma/\rho_0 SC_L$$

$$d\lambda_v/d\gamma = \lambda_v C_D/C_L$$

Solving the above state and costate equations,

$$V\lambda_v = V_n \lambda_{vn} = \text{constant}$$

$$\lambda_H \exp(\beta h) = \lambda_{Hn} \exp(\beta H_n)$$

The boundary conditions for the adjoint variables are

$$\lambda_v(\gamma = \gamma_f) = \left.\frac{\partial J}{\partial V}\right|_{\gamma = \gamma_f} = -1$$

$$\lambda_H(\gamma = \gamma_f) = \left.\frac{\partial J}{\partial H}\right|_{\gamma = \gamma_f} = 0$$

Solving the above, we get

$$\lambda_H = 0$$

Then the Hamiltonian reduces to

$$\mathcal{H} = -\lambda_v V[C_{DO}/C_L + KC_L]$$

The optimal control is then given by

$$\frac{\partial \mathcal{H}}{\partial C_L} = 0$$

leading to

$$C_{LO} = \sqrt{C_{DO}/K} = C_{LE}$$

where, C_{LE} is the lift coefficient for maximum lift-to-drag ratio $(L/D)_{max} = E = 1/2\sqrt{KC_{DO}}$. With the above optimal control in (26), and noting that $\gamma_n = 0$, we solve for the velocity and altitude as

$$V(\gamma) = V_n \exp(-\gamma/E)$$

$$H = \ln\left[\exp(-\beta H_n) + \frac{2\beta m_n}{\rho_0 SC_{LO}}(\cos\gamma - 1)\right]^{-1/\beta}$$

For small γ, it reduces to

$$H = \ln\left[\exp(-\beta H_n) - \frac{\beta m_n}{\rho_0 SC_{LO}}\gamma^2\right]^{-1/\beta}$$

At the end of the ascent mode, $\gamma = \gamma_f$ and $H = H_f$. Then the above relation becomes

$$H_f = \ln\left[\exp(-\beta H_n) - \frac{\beta m_n}{\rho_0 SC_{LO}}\gamma_f^2\right]^{-1/\beta}$$

Using earlier relation for H_j and noting that $H_e = H_f$ and $H_j = H_n$, we get

$$\gamma_f = -\gamma_e\sqrt{m_j/m_n} \tag{27}$$

The above relation shows that at the end of the atmospheric phase, the vehicle has to leave the atmosphere with a positive flight path angle higher in magnitude to that of the entry flight path angle. This is due to the fact that the mass m_n at the beginning of the ascent mode (or end of the cruise mode) is less than the mass m_j at the end of the descent mode (or beginning of the cruise mode).

5.5 Boost and Reorbit Phase

During the atmospheric flight, the vehicle performs the desired plane change and dissipates some energy due to atmospheric drag. Therefore, a second impulse is required to boost the vehicle back to orbital altitude Fig. 1. The vehicle exits the atmosphere at point F, with a velocity V_f and flight path angle γ_f. The additional impulse ΔV_b, required at the exit point F for boosting into an elliptic orbit with apogee radius R_c and the reorbit impulse ΔV_c required to insert the vehicle into a circular orbit at point C, are obtained by using the principle of conservation of energy and angular momentum at the exit point F, and the circularization point C. Thus, we have [13],

$$(V_f + \Delta V_b)^2/2 - \mu/R_a = (V_c - \Delta V_c)^2/2 - \mu/R_c$$

$$(V_f + \Delta V_b)R_a \cos\gamma_f = R_c(V_c - \Delta V_c)$$

Solving for ΔV_b and ΔV_c,

$$\Delta V_b = \sqrt{2\mu(1/R_a - 1/R_c)/\left[1 - (R_a/R_c)^2\cos^2\gamma_f\right]} - V_f$$

$$\Delta V_c = \sqrt{\mu/R_c} - \sqrt{2\mu(1/R_a - 1/R_c)/\left[(R_c/R_a^2)/\cos^2\gamma_f - 1\right]}$$

Finally, the vehicle is in a circular orbit (of radius R_c) moving with the velocity $V_c = \sqrt{\mu/R_c}$.

5.6 Numerical Simulation

The following set of data is used for a typical orbital maneuvering research vehicle [5-8].

Orbital Data

Altitude of HEO, H_d = 115,000 m

Altitude of LEO, H_c = 115,000 m

Altitude of atmospheric boundary, H_a = 110,000 m

Radius of Earth, R_E = 6,356,766 m

Acceleration due to gravity, g_0 = 9.80665 m/sec^2

Atmospheric density at sea level, ρ_0 = 1.225 kg/m^3

Gravitational constant of Earth, μ = 3.986x10^4 m^3/sec^2

Inverse atmospheric scale height, β = 1/7280 m^{-1}

Vehicle Data

Initial mass, m_j = 4760 kg

Propellant available for cruise, $(m_j - m_n)$ = 1810 kg

Final mass, m_n = 2950 kg

Aerodynamic reference area, S = 11.613 m^2

Specific fuel impulse, I_{sp} = 290 sec

The aerodynamic characteristics are described in terms of the angle of attack α as

$$C_L = -2.068686996\alpha^3 + 2.943200144\alpha^2 + 0.080347684\alpha + 0.031320026$$

$$C_D = 0.267339707\alpha^3 + 1.814473159\alpha^2 - 0.389985867\alpha + 0.068372034$$

Fig. 4 shows the variations of the lift and drag coefficients C_L, and C_D and Fig. 5 shows the variations of the lift-to-drag ratio E, and the aeropropulsive efficiency E_p as a function of the angle of attack α. The maximum lift-to-drag ratio of 2.3149 occurs at the angle of attack of 13 degrees.

Deorbit Phase

Initially, the vehicle is at a HEO altitude H_d of 115 km orbiting with a circular velocity V_d of 7847.97 m/sec. A deorbit impulse ΔV_d of 518.99 m/sec puts the vehicle in an elliptic orbit to intersect the atmospheric boundary at an altitude H_a of 110 km. At the atmospheric entry point, the velocity V_e is 7334.17 m/sec and the flight path angle γ_e is -0.77 deg.

Atmospheric Phase: Deorbit Mode

During the descent mode of atmospheric phase, the vehicle descends from an altitude H_a of 110,000 m at a velocity V_e of 7334.17 m/sec to a cruise altitude H_j of 72,521 m and the cruise velocity V_j of 7291.17 m/sec. During

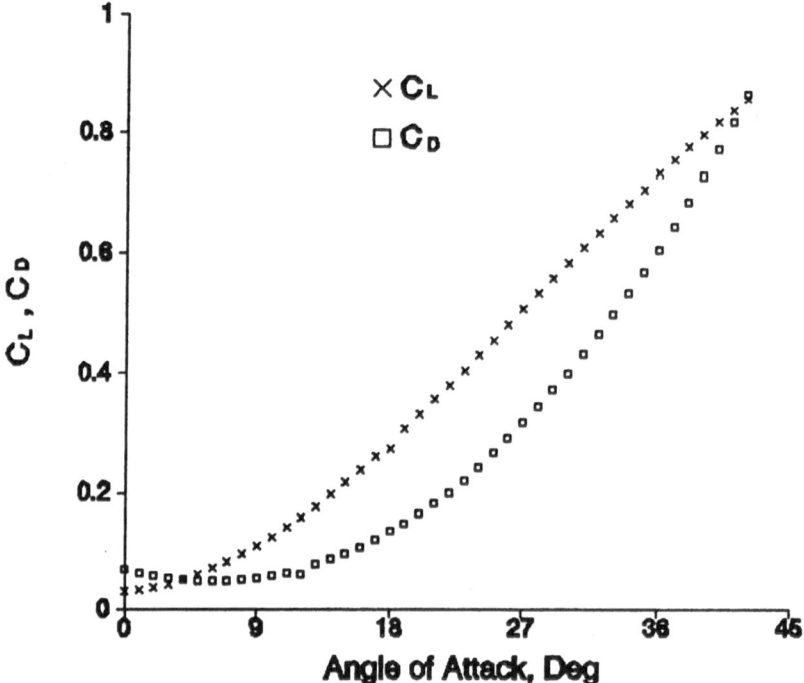

Fig. 4 Lift and Drag Coefficients

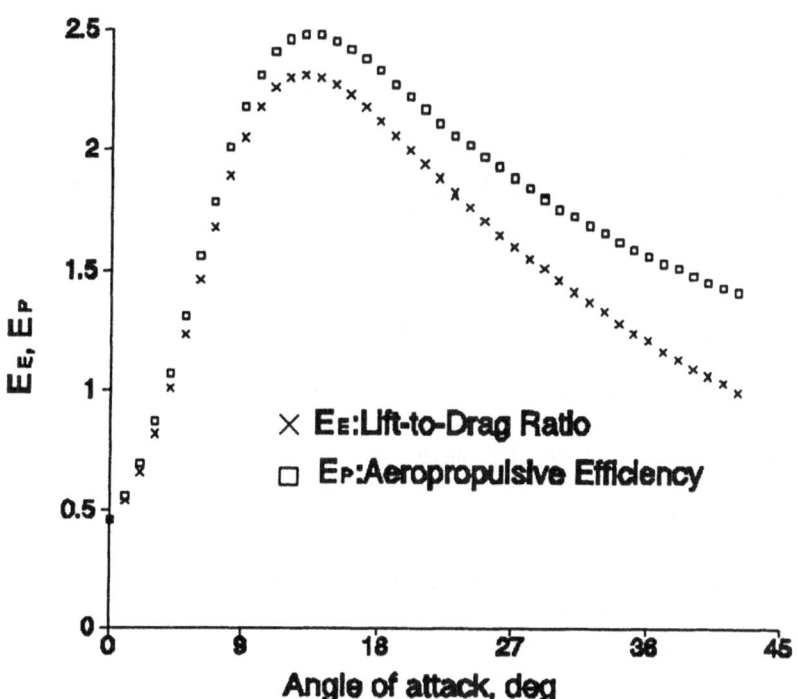

Fig. 5 Lift-Drag Ratio and Aeropropulsive Efficiency

this time, the vehicle is maintained at an angle of attack of 13 degrees corresponding to the lift for maximum lift-to-drag ratio.

Cruise Mode: Bank Angle Control

The cruise mode is analyzed using bank angle control or thrust control. With bank angle control, for the same fuel consumption and a given L/D (or angle of attack α), the optimum thrust angle as obtained from (17) is represented in Fig. 6. With this optimum thrust angle, the corresponding heading angle is obtained from (15) and is shown in Fig. 7. It is seen that at these constant cruising conditions (of altitude of 72521 m, velocity of

7291.7 m/sec, and thrust of 6108.9 Nw), the maximum heading and hence maximum inclination of 18.56 degrees is achieved with a thrust angle of 44.72 degrees and at a higher angle of attack of 24 degrees rather than at the angle of attack of 13 degrees corresponding to maximum L/D [5-8].

Cruise Mode: Thrust Control

With thrust control, for the same fuel consumption and a given L/D (or angle of attack α), the optimum bank angle as obtained from (23) is shown in Fig. 8. With this optimum bank angle, the corresponding heading angle is obtained from (22) and is shown in Fig. 9. It is seen that at these cruising conditions, the maximum heading angle of 17.7 degrees is achieved with a bank angle of 51.9 degrees and at a higher angle of attack of 20 degrees rather than at the angle of attack of 13 degrees corresponding to maximum L/D.

The comparison of maximum heading angle as a function of angle of attack for both the control strategies shown in Fig. 10, indicates the superiority of bank control over thrust control. It is to be noted that the heading achieved depends on the type of control used, cruise conditions, and the angle of attack.

Atmospheric Phase: Ascent Mode

At the end of cruise mode, the vehicle ascends to the atmospheric boundary with a constant angle of attack of 13 degrees corresponding to maximum lift-to-drag ratio. At the end of the ascent mode, the exit velocity V_f is 7238.1 m/sec, the flight path angle γ_f as given by (27) is 0.9781 deg.

For the atmospheric phase with bank angle control, the total time solutions for altitude, velocity, flight path angle, heading angle, and heating rate are shown in Fig. 11. Similarly, total solutions for thrust control are shown in Fig. 12. The heating rate is computed from

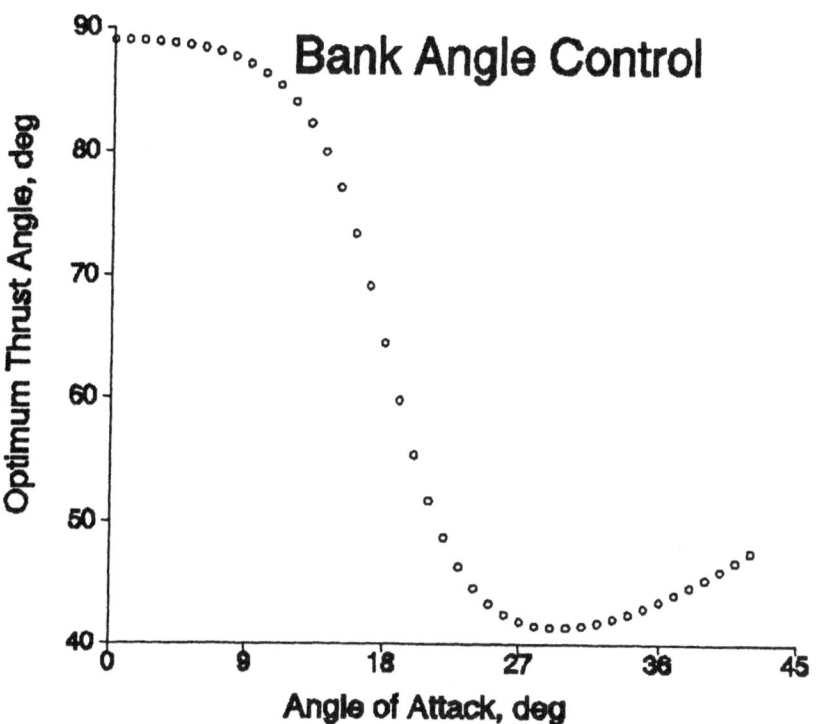

Fig. 6 Optimal Thrust Angle with Bank Angle Control

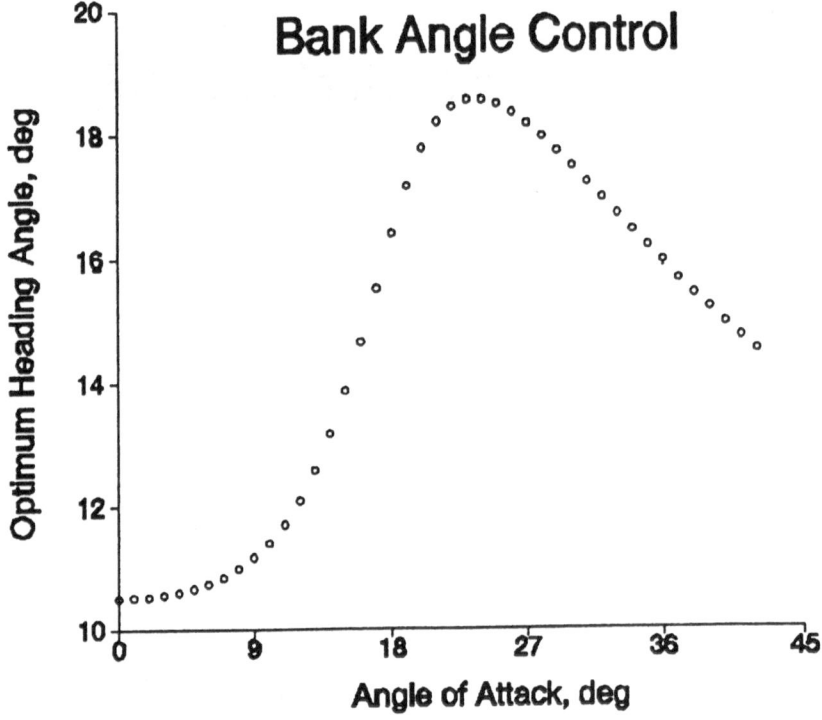

Fig. 7   Optimal Heading Angle with Bank Angle Control

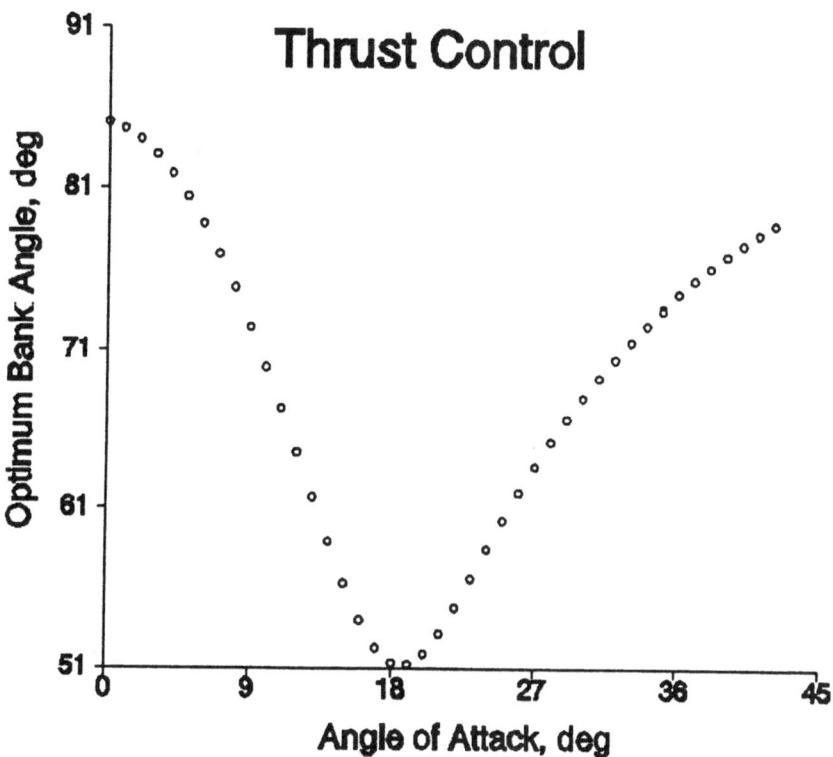

Fig. 8 Optimal Bank Angle with Thrust Control

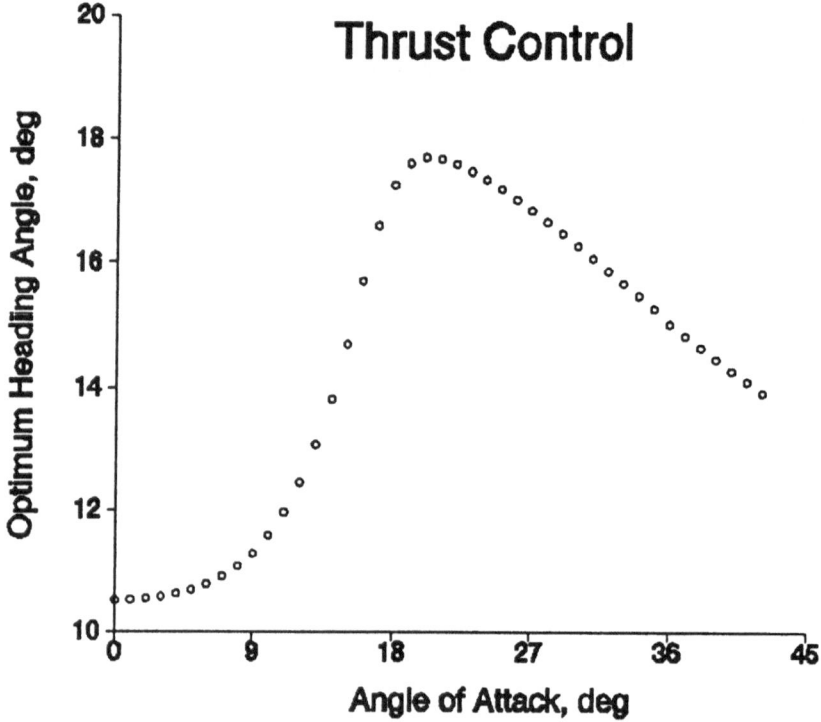

Fig. 9   Optimal Heading Angle with Thrust Control

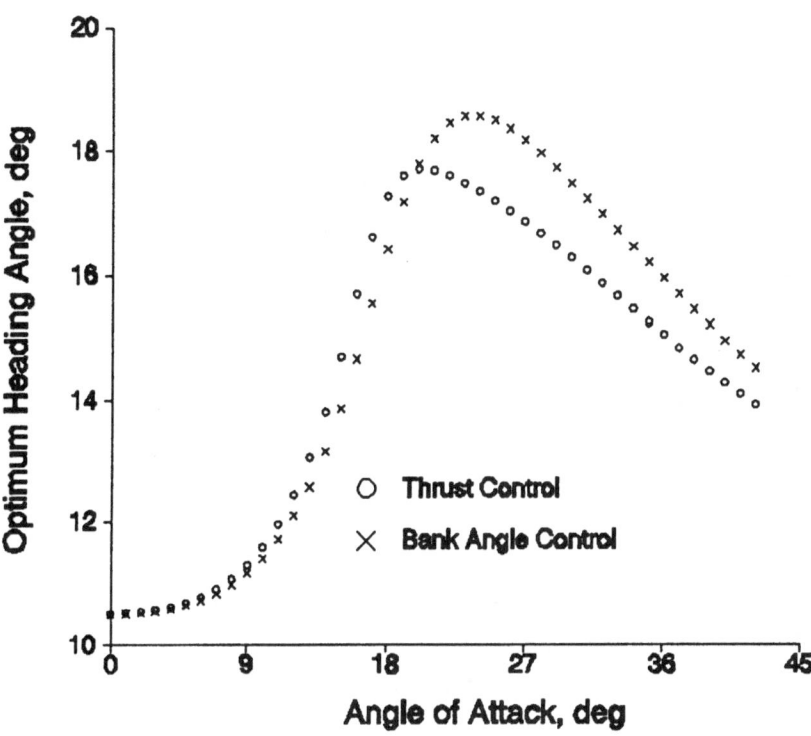

Fig. 10 Optimal Heading Angle for Bank Angle Control and Thrust Control

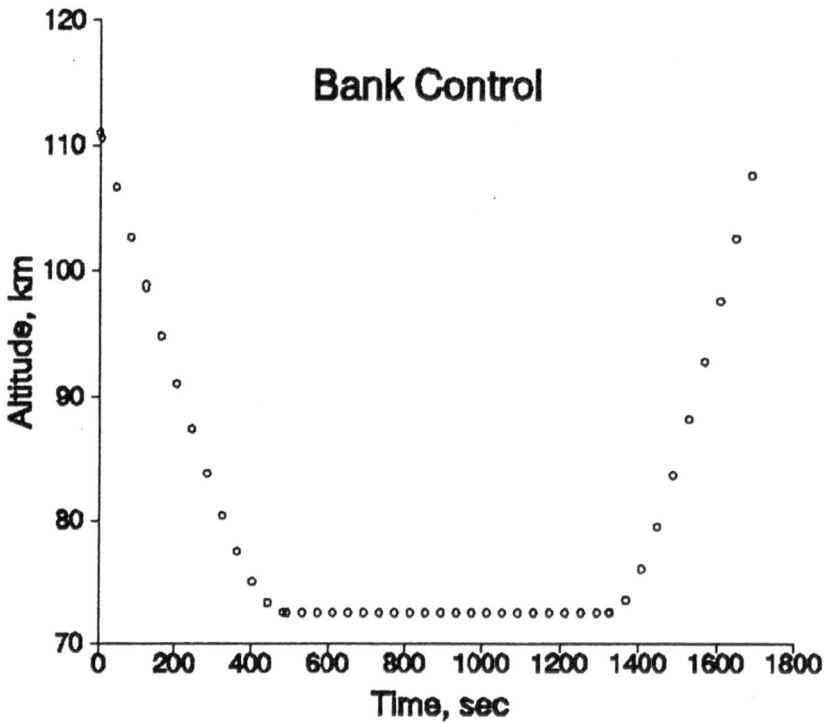

Fig. 11(a)   Altitude with Bank Angle Control

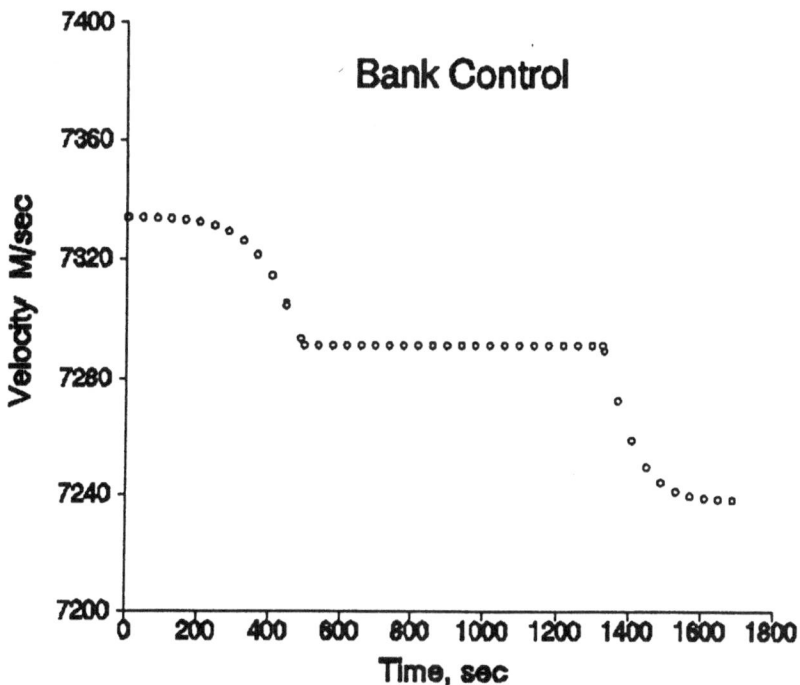

Fig. 11(b)   Velocity with Bank Angle Control

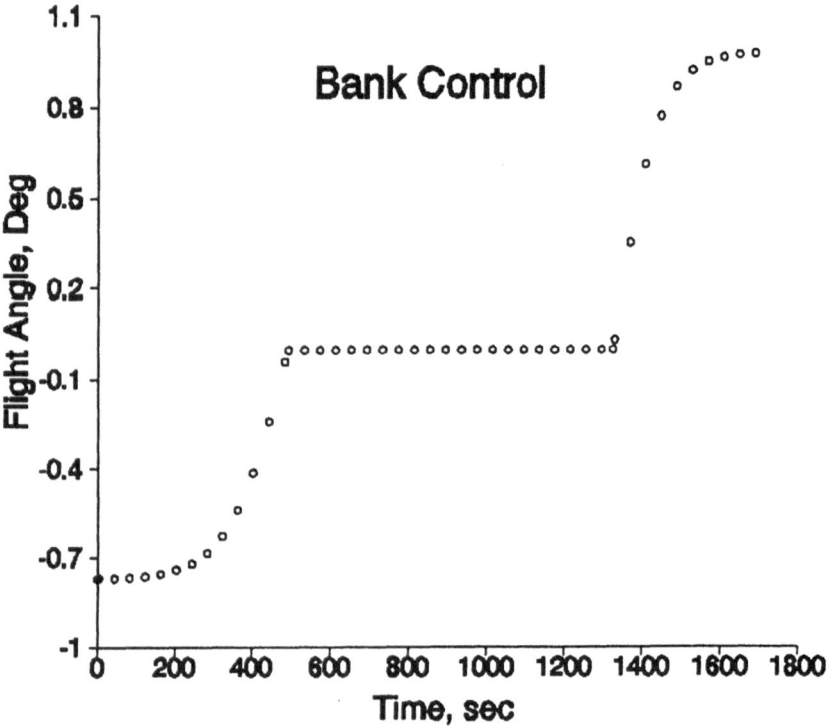

Fig. 11(c) Flight Path Angle with Bank Angle Control

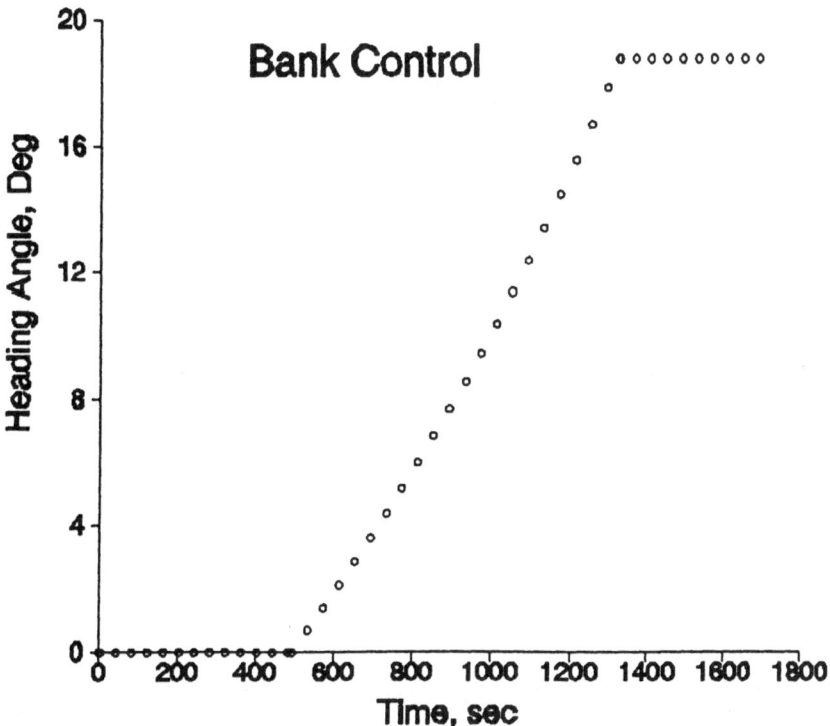

Fig. 11(d)   Heading Angle with Bank Angle Control

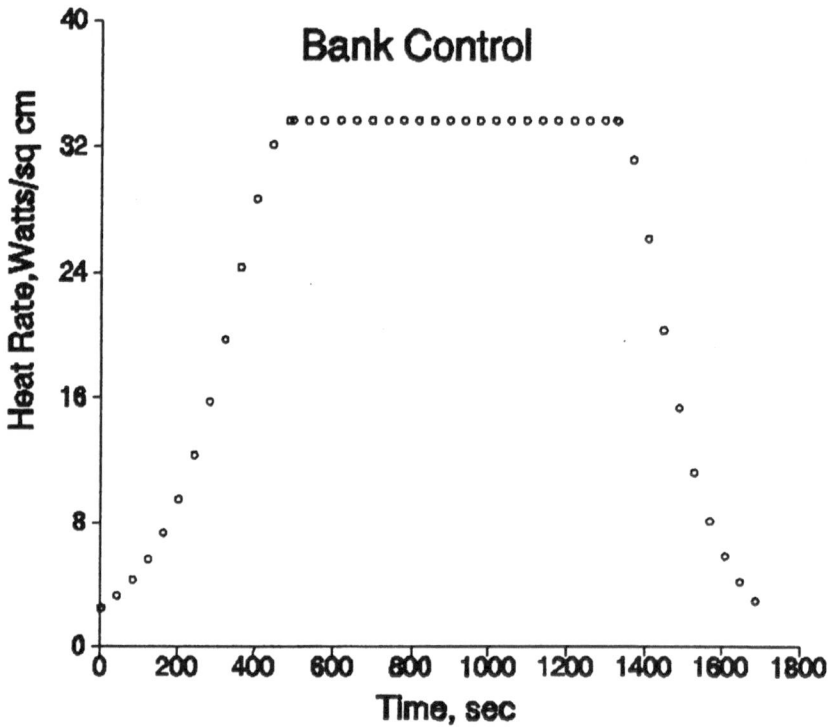

Fig. 11(e)   Heat Rate with Bank Angle Control

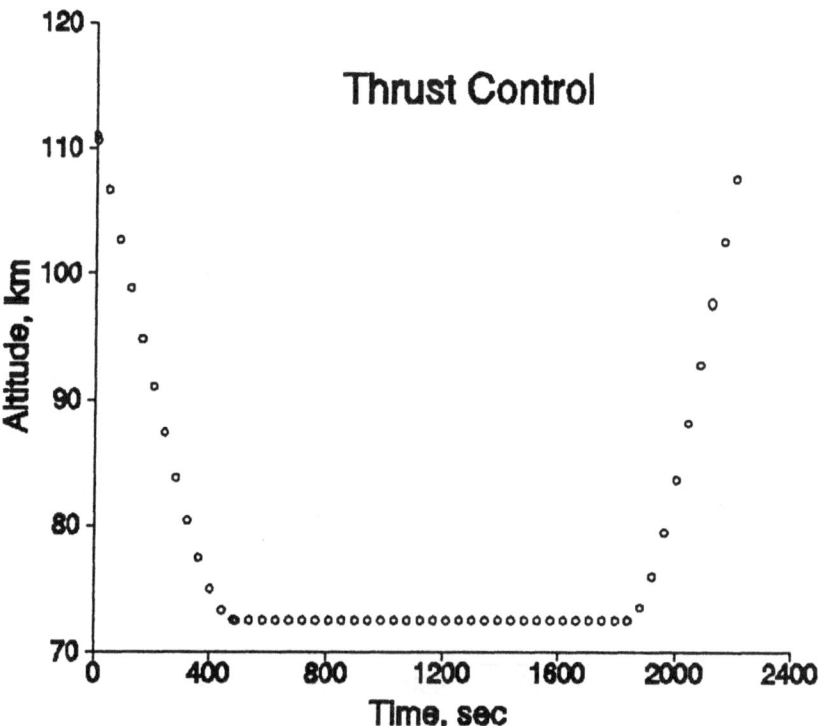

Fig. 12(a)   Altitude with Thrust Control

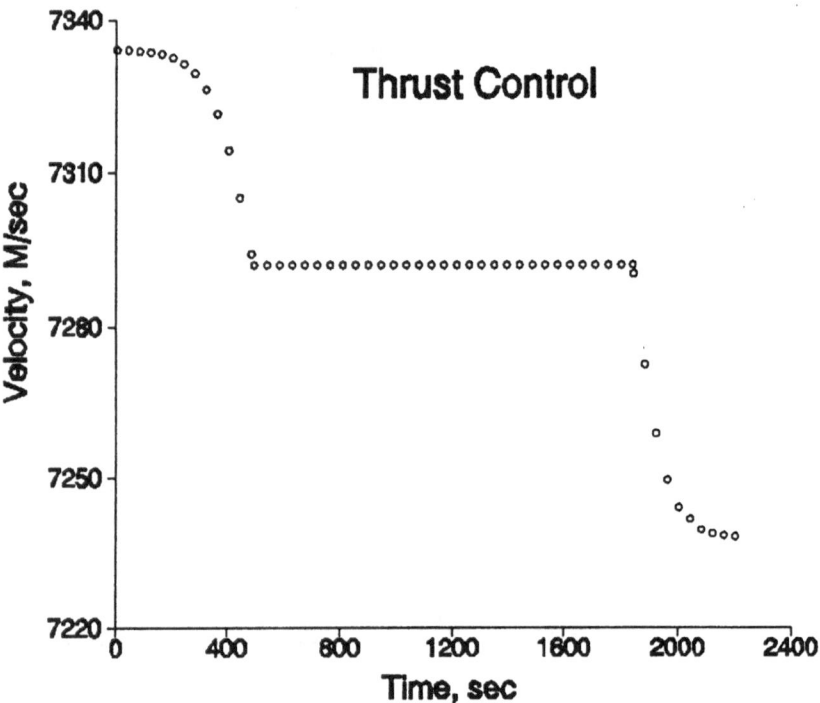

Fig. 12(b)   Velocity with Thrust Control

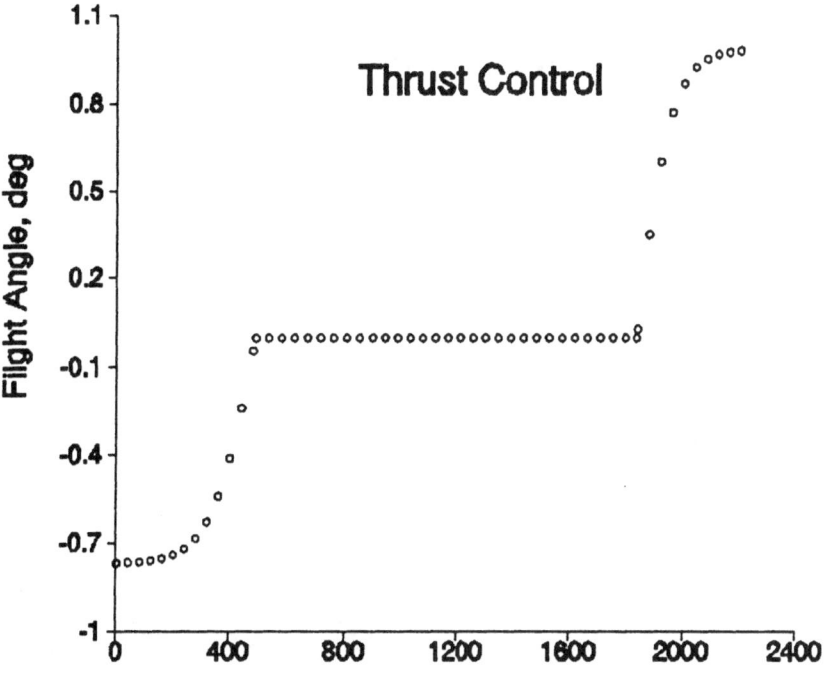

Fig. 12(c) Flight Path Angle with Thrust Control

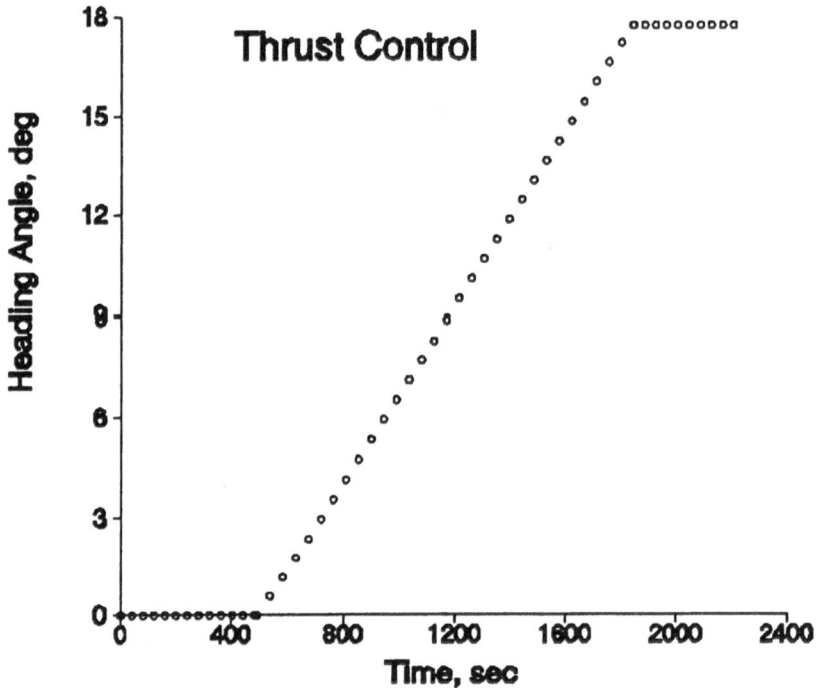

Fig. 12(d)   Heading Angle with Thrust Control

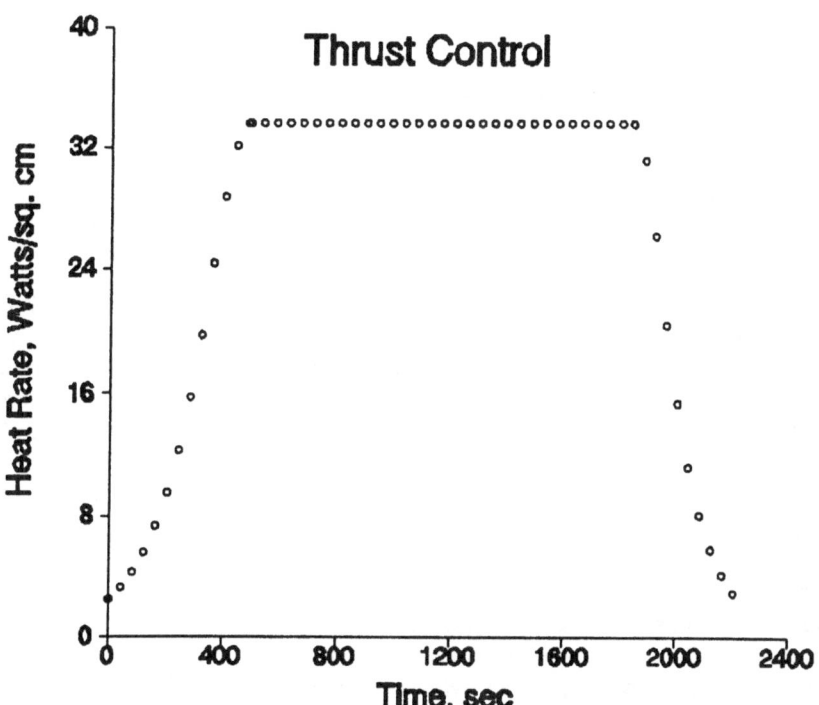

Fig. 12(e) Heat Rate with Thrust Control

$$\dot{Q} = 3.08 \times 10^{-4} \rho_k^{1/2} V_k^{3.08} \text{ Watts/cm}^2$$

where, ρ_k is expressed in kg/km^3, and V_k is expressed in km/sec.

Reorbit Phase

At the end of the atmospheric phase, a boost impulse ΔV_b of 380 m/sec is executed to bring the vehicle to its original altitude H_c of 110 km. At this time, once again a circularizing impulse ΔV_c of 247.97 m/sec is imparted to finally put the vehicle in circular orbit.

5.7 Concluding Remarks

This Chapter has addressed the synergistic plane change problem in connection with orbital transfer employing aeroassist technology. The mission involved transfer from high Earth orbit to low Earth orbit with plane change being performed within the atmosphere. The complete mission consisted of a deorbit phase, an atmospheric phase, and finally a reorbit phase. The atmospheric maneuver was composed of an entry mode, a cruise mode, and finally an exit mode. The descent and ascent modes have been analyzed using flight path angle as an independent variable for maximizing the cruise and exit velocities with a constraint on the minimum cruise altitude. During the cruise mode, constant altitude and velocity were maintained by means of bank angle control with constant thrust or thrust control with constant bank angle. Conditions have been obtained for maximizing the heading angle. Under given cruising conditions, the maximum heading angle has been achieved with an angle of attack higher than that corresponding to the maximum lift-to-drag ratio. Comparison between the two control strategies has shown the superiority of bank control over thrust control in terms of the maximum achievable heading angle.

Nomenclature

C_D : drag coefficient

C_{DO}: drag coefficient at zero lift

C_L : lift coefficient

C_{LR}: lift coefficient at maximum lift-to-drag ratio

D : drag force

E : maximum value of lift-to-drag ratio

E_p : aeropropulsive efficiency
g : gravitational acceleration
H : altitude
$\mathcal{H}$: Hamiltonian
I_{sp} : specific fuel consumption
i : inclination
J : performance index
K : induced drag factor
L : lift force
m : vehicle mass
Q : Heating rate
R : distance from Earth center to vehicle
R_E : radius of Earth
S : aerodynamic reference area
T : thrust
t : time
V : velocity
α : angle of attack
β : inverse atmospheric scale height
γ : flight path angle
δ : normalized density
ζ = normalized lift coefficient
η : thrust angle
θ : down range angle
λ : costate (Lagrange) variable
μ : gravitational constant of Earth
ρ : density
σ : bank angle
ϕ : cross range angle
ψ : heading angle
ΔV : characteristic velocity

Subscripts

a : atmospheric boundary
c : circularization at LEO
d : deorbit at HEO
e : entry to atmosphere
f : exit from atmosphere
j : beginning of aerocruise
n : end of aerocruise
s : surface level

References

[1] G. D. Walberg, "A Survey of Aeroassisted Orbital Transfer," *Journal of Spacecraft*, 22, 3-18, 1985

[2] F. S. Nyland, "Considerations of applying continuous thrust during synergistic plane change," *AAS 68-121, AAS/AIAA Astrodynamic Specialists Conference*, 1-9, 1968.

[3] E. Cuadra, and P. D. Arthur, "Orbit plane change by external burning aerocruise," *Journal of Spacecraft & Rockets*, 3, 347-352, 1966.

[4] J. S. Clauss Jr., and R. D. Yeatman, "Effect of heating restraints on aeroglide and aerocruise synergistic maneuver performance," Journal of Spacecraft and Rockets, 4, 1107-1109, 1967.

[5] H. Ikawa, and T. F. Rudiger, "Synergistic maneuvering of a winged spacecraft for orbital plane change," *Journal Spacecraft & Rockets*, 19, 513-520, 1982.

[6] R. T. Cervisi, "Analytical solutions of a cruising plane change maneuver," *Journal of Spacecraft & Rockets*, 22, 134-140, 1985.

[7] J. A. Blissit Jr., *An adaptive guidance algorithm for an aerodynamically assisted orbital plane change maneuver*, M.S. Thesis, Draper Labs., Cambridge, MA, June 1986.

[8] K. D. Mease, J. Y. Lee, and N. X. Vinh, "Orbital changes during hypersonic aerocruise," *The Journal of Astronautical Sciences*, 36,103-137, 1985.

[9] D. S. Naidu, J. L. Hibey, and C. Charalambous, "Fuel-optimal trajectories for aeroassisted coplanar orbital transfer problem," *IEEE Transactions on Aerospace and Electronic Systems*, 26, 374-381, 1990.

[10] D. S. Naidu, "Fuel-optimal trajectories for aeroassisted orbital transfer with plane change," *IEEE Transactions on Aerospace and Electronic Systems*, 26, 1991.

[11] J. P. Marec, *Optimal Space Trajectories*, Elsevier Scientific Publishing Company, Amsterdam, 1979.

[12] N. X. Vinh, *Optimal Trajectories in Atmospheric Flight.* Elsevier Scientific Publishing Co., Amsterdam, 1981.

13. L. Beiner, "A simplified three-dimensional altitude constrained skip maneuver for maximum orbit plane change," *Israel Journal of Technology*, 13, 82-88, 1975.

14. R. N. Bell, and W. L. Hankey Jr., *Applications of aerodynamic lift in accomplishing orbital plane change*, Air Force Systems Command, Wright-Patterson Air Force BAse, Ohio, ASD-TDR-63, 693, Sept. 1963.

****

CHAPTER 6

OPTIMAL GUIDANCE FOR ORBITAL TRANSFER

6.1 Introduction

In establishing a permanent space station, the aeroassist technology is an important ingredient in space transportation system. The use of aeroassisted maneuvers to affect a transfer from high Earth orbit (HEO) to low Earth orbit (LEO) has been recommended to provide high performance leverage to future space transportation systems [1]. The aeroassisted orbital transfer vehicle (AOTV), on its return journey from HEO, dissipates orbital energy through atmospheric drag to slow down to LEO velocity. Thus, the basic idea is to employ a hybrid combination of propulsive maneuvers in space and aerodynamic maneuvers in sensible atmosphere. Within the atmosphere, the trajectory control is achieved by means of lift and bank angle modulations. Hence, this type of flight with a combination of propulsive and nonpropulsive maneuvers, is also called synergetic maneuver or space flight.

6.2 Neighboring Guidance

Guidance is the determination of a strategy for following a nominal flight in the presence of off-nominal conditions, wind disturbances, and navigation uncertainties [2-5]. There are two fundamentally different approaches for guidance of spacecraft through the atmosphere.

6.2.1 Predicted Guidance

Here, predict future trajectories at the required point of time by either fast computation or use of approximate closed form solutions. In prediction using fast computation, the governing equations are solved by an

This Chapter is based on (i) D. S. Naidu, "Neighboring optimal guidance for aeroassisted noncoplanar orbital transfer," Originally presented at the AIAA Atmospheric Flight Mechanics Conference, Aug. 12-14, 1991, New Orleans, Louisiana. The permission given by AIAA is hereby acknowledged. (ii) D. S. Naidu, "Neighboring optimal guidance for aeroassisted noncoplanar orbital transfer," Intl. Journal of Systems Science, 24, 563-575, March 1993. The permission given by Taylor & Francis is hereby acknowledged.

air-borne computer to determine all possible future trajectories. The main advantage of this type of prediction is the ability to handle any possible flight condition and accommodate large off-nominal conditions. The principal objection to the system is the requirement of large on board computer. On the other hand, for the on board prediction of trajectories, the closed form solutions are very simple and convenient for implementation. In most of the cases, the closed form solutions are obtained based on approximate techniques. Thus, the trajectories are obtained typically for constant altitude, constant deceleration, and equilibrium glide paths. Guidance using approximate closed-form solutions has the disadvantage of not having the flexibility to handle enough off-nominal conditions.

6.2.2 Nominal Guidance

In this scheme, the nominal trajectories are precomputed on ground taking into all the aspects of constraints with optimization and stored on board. During the flight, the difference between the measured values and nominal (stored) values is used in guiding the vehicle onto the nominal trajectory (path controller) or generate a new trajectory to reach the destination (terminal controller). Here, the nominal trajectory being fixed on the ground, does not take into account any contingencies arising during the flight.

In a typical guidance scheme, the final steering command is generated as the sum of two components, an open-loop actuating (control) signal yielding the desired vehicle trajectory in the absence of external disturbances, and a linear feedback regulating signal which reduces the system sensitivity to unwanted influences on the vehicle. It is well known that the closed-loop system is stable about the nominal trajectory and has additional desirable features.

This Chapter, addresses the fuel-optimal control problem arising in noncoplanar orbital transfer employing aeroassist technology. The maneuver involves the transfer from HEO to LEO with a prescribed plane change and at the same time minimization of the fuel consumption. It is known that the change in velocity, also called the characteristic velocity, is a convenient parameter to measure the fuel consumption. For minimum-fuel maneuver, the objective is then to minimize the total characteristic velocity for deorbit, boost, and reorbit (or circularization). The corresponding optimal (nominal) trajectory and control for the atmospheric maneuver are obtained. The linearization is performed around the nominal condition and the resulting model is fitted into the framework of linear quadratic regulator (LQR) theory. Instead of using time, one of the state variables is employed as an independent variable, thus avoiding any control required due to irrelevant timing errors. Also, the elimination of time as an independent variable carries with it the advantage of order reduction for the system. The choice of weighting matrices in the performance index is made by combining a heuristic method and optimal modal control approach. The feedback control law is obtained to suppress the perturbations from the nominal condition. The results are shown for a typical AOTV.

6.3 Formulation Of The Problem

The basic equations for orbital transfer from HEO to LEO, are those for deorbit, aeroassist (or atmospheric flight), boost and circularization (or reorbit). But, for guidance and control purposes, consider only the atmospheric flight during which phase the vehicle needs to be controlled by aerodynamic lift and bank angle to achieve the necessary velocity reduction and the plane change.

Consider a vehicle moving about a nonrotating spherical planet. The atmosphere surrounding the planet is assumed to be at rest, and the central gravitational field obeys the usual inverse square law. The equations of motion for the vehicle are given for kinematics as [6],

$$\frac{d\theta}{dt} = V\cos\gamma\cos\psi/R\cos\phi$$

$$\frac{d\phi}{dt} = V\cos\gamma\sin\psi/R$$

$$\frac{dR}{dt} = V\sin\gamma$$

The force equations are

$$m\frac{dV}{dt} = -D - mg\sin\gamma$$

$$mV\frac{d\gamma}{dt} = L\cos\sigma + m(V^2/R - g)\cos\gamma$$

$$mV\frac{d\psi}{dt} = L\sin\sigma/\cos\gamma - (mV^2/R)\cos\gamma\cos\psi\tan\phi \tag{1}$$

where,

$$L = C_L\rho SV^2/2; \quad D = C_D\rho SV^2/2; \quad C_D = C_{DO} + KC_L^2$$

$$g = \mu/R^2; \quad R = H + R_E; \quad \rho = \rho_s\exp(-H\beta)$$

Using the normalized variables,

$$\tau = t/\sqrt{R_a^3/\mu}; \quad v = V/\sqrt{\mu/R_a}$$

and the dimensionless constants,

$$h = H/H_a; \quad b = R_a/H_a; \quad \delta = \rho/\rho_s = \exp(-h\beta H_a)$$

$$\eta = C_L/C_{LR}; \quad C_{LR} = \sqrt{C_{DO}/K}$$

in (1) yields the normalized form as

$$\frac{d\theta}{d\tau} = \frac{bv\cos\gamma\cos\psi}{(b-1+h)\cos\phi}$$

$$\frac{d\phi}{d\tau} = \frac{bv\cos\gamma\sin\psi}{(b-1+h)}$$

$$\frac{dh}{d\tau} = bv\sin\gamma$$

$$\frac{dv}{d\tau} = -A_1 b(1+\eta^2)\delta v^2 - \frac{b^2\sin\gamma}{(b-1+h)^2}$$

$$\frac{d\gamma}{d\tau} = A_2 b\eta\delta v\cos\sigma + \frac{bv\cos\gamma}{(b-1+h)} - \frac{b^2\cos\gamma}{(b-1+h)^2 v}$$

$$\frac{d\psi}{d\tau} = \frac{A_2 b\delta\eta v\sin\sigma}{\cos\gamma} - \frac{bv\cos\gamma\cos\psi\tan\phi}{(b-1+h)} \qquad (2)$$

where, $A_1 = C_{DO}Sp_s H_a/2m; \quad A_2 = C_{LR}Sp_s H_a/2m$

Taking the performance index as the sum of the characteristic velocities for deorbit, boost, and recircularization, the above fuel-optimal problem is solved by using a multiple shooting method. This is the nominal (optimal) solution to be used in obtaining the guidance scheme [7].

However, under the actual flight conditions, the initial conditions may not agree with those assumed for generating the nominal trajectory and more important the vehicle may not follow the nominal trajectory due to various disturbances. Hence, rather than compute a new nominal trajectory each time some disturbance is encountered, it is proposed to implement a closed-loop control of the vehicle to compensate for the deviations from the nominal condition.

6.3.1 Selection of Independent Variable

In a typical guidance problem, the vehicle is required to follow a specified (nominal) trajectory in three-dimensional space. Suppose, the actual vehicle at a particular time is on the nominal trajectory, but much earlier than required or reached the desired point at incorrect time. The exact time at which the vehicle reaches the various points on the trajectory may be of no concern. Thus, if the actual trajectory satisfies the spatial relationship of the desired (nominal) trajectory, but the vehicle travelled more slowly or more quickly, the regulator (designed with time as independent variable) would sense these timing errors as spatial (state) errors along the trajectory, although, in fact, these are errors of no relevance to the objective of the guidance [2-4]. Based on these spurious errors, the regulator tends to "over control" the system and thus wastes considerable amounts of control effort. Because time is used as the independent variable in the regulator design, not only does the regulator try to follow the desired path, but it tries to force the system to follow at a particular rate.

The LQ regulator can be made significantly more robust by the simple technique of using a trajectory (state) variable instead of time as the independent variable. Also, if one is concerned only with the relationship that should exist between the system trajectory variables on the desired path, one of the state variables could be used in place of time as the independent variable. The main appeal for this change of the independent variable lies in the fact that inherently state dependent events become time-dependent events in an indirect manner. Another obvious advantage of eliminating time as the independent variable is simply the reduction in the order of the mathematical model of the system.

In the present plane change problem, the down range θ is monotonic and is a good candidate for an independent variable. Thus, the system of equations (2) with time as independent variable is converted into the following system of equations with down range θ as the independent variable.

$$\frac{dh}{d\theta} = (b-1+h)\tan\gamma\cos\phi/\cos\psi$$

$$\frac{dv}{d\theta} = -A_1(1+\eta^2)\delta v(b-1+h)\cos\phi/\cos\gamma\cos\psi$$

$$\frac{d\gamma}{d\theta} = A_2\delta\eta(b-1+h)\cos\sigma\cos\phi/\cos\gamma\cos\psi$$

$$\frac{d\phi}{d\theta} = \cos\phi\tan\psi$$

$$\frac{d\psi}{d\theta} = A_2 \delta\eta(b\text{-}1\text{+}h)\sin\sigma\cos\phi/(\cos\gamma)^2\cos\psi \tag{3}$$

where, δ is a function of h, and the controls are normalized lift coefficient η and bank angle σ. In the above equations, a valid assumption that the gravity and centrifugal forces are negligible compared to either lift force or drag force is used in order to get a simplified form of equations. Also, note that the order of the system (3) is now only five compared to the sixth order system (2).

6.3.2 Linearization

The nonlinear equations of motion (3) are represented in a compact form as

$$\dot{x} = f(x,u)$$

where, $\dot{x} = dx/d\theta$, the state vector $x^T = [h,v,\gamma,\phi,\psi]$ and the control vector $u^T = [\eta,\sigma]$. Let the system (2) represent the nominal (optimum) trajectory and control which we are interested in flying. The optimization results in an open-loop implementation. Unfortunately, the open-loop guidance tends to be very sensitive to external disturbances and the vehicle parameter changes. In an actual situation, the system equations may differ from reality due to various assumptions made in deriving them. Consider the perturbations as

$$x = x_o + \delta x; \quad u = u_o + \delta u$$

where, $[x_o, u_o]$, and $[x, u]$ represent nominal and perturbed conditions respectively, and $[\delta x, \delta u]$ represent the deviations from the nominal condition.

If we aim at the guidance laws that meet the performance specifications in the presence of off-nominal conditions, it is natural to seek the principles of linearity and feedback. Then, it is well known that the perturbed system satisfies the linear system [8]

$$\delta\dot{x} = A\delta x + B\delta u \tag{4}$$

where, $A = \partial f/\partial x$, and $B = \partial f/\partial u$, are evaluated along the nominal trajectory. Although the matrices A and B can be evaluated for all values of interest, it may sometimes be necessary to keep A and B constant between the intervals of updating. The errors due to this quantization procedure are assumed to be small enough so that no drastically different changes occur in the performance. Keeping A and B constant also enables us to use the time-invariant version of the Riccati equation in the regulator problem.

6.4 Neighboring Optimal Guidance

The basic regulator synthesis problem is that of selecting a control policy which generates δu in such a way that $[\delta x, \delta u]$ are made as small as possible [8]. The performance index is

$$J = \frac{1}{2} \int_0^\infty [\delta x^T Q \delta x + \delta u^T R \delta u] d\theta$$

where, the first and second terms under the integral represent the penalty from the nominal state and control vectors respectively. The matrix R must be positive definite, while Q must be positive semi definite. These weighting matrices are usually symmetric and diagonal to minimize the number of parameters to be chosen.

It is shown that the omission of second order terms in getting the linearized equation is justified by selecting the performance index with quadratic character. Using the minimization procedure, the feedback law is obtained as

$$\delta u = - F \delta x$$

Where $F = R^{-1} B^T P$ and P is the positive definite, symmetric solution of the matrix algebraic Riccati equation (ARE)

$$PA + A^T P - PBR^{-1}B^T P + Q = 0$$

The linear optimal control theory results in a feedback law which computes corrections to the nominal commands as shown in Fig. 1.

Desirable Features of LQR Guidance

(i) The closed-loop optimal control is linear.

(ii) The control is easy for implementation or mechanization.

(iii) The closed-loop system using the optimal control is asymptotically stable.

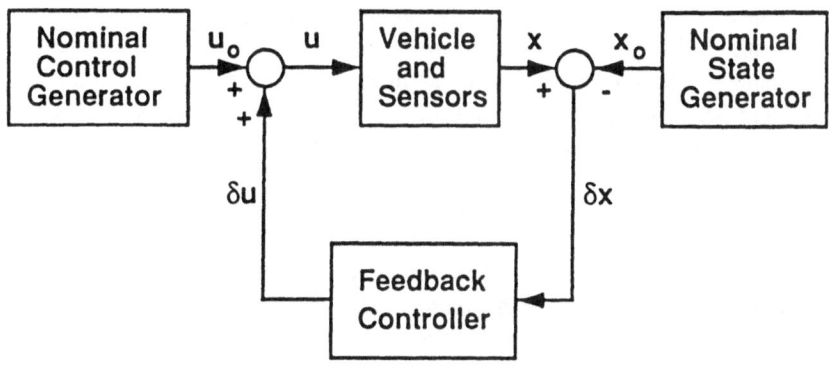

Fig. 1 Neighboring Optimal Guidance

(iv) The closed-loop control is robust in the sense that, if the system contains additional white-noise terms which model turbulence, atmospheric inhomogeneities, and unmodelled high-frequency vehicle dynamics, the closed-loop control will still give the control policy which minimizes the expected value of the performance index.

(v) The closed-loop system possesses attractive sensitivity properties with respect to variations in the elements of A and B.

(vi) The guidance scheme is off-line, in the sense that the feedback coefficient matrix F $(= R^{-1}B^{T}P)$ is calculated in advance and stored for implementation on board.

6.5 Selection Of Weighting Matrices

Guidance requirements during the atmospheric maneuver demand good nominal path following, and small errors under a variety of perturbations and off-nominal conditions. The designer must be able to come up with a set of weighting matrices Q and R that satisfy these specifications. In general, there is no simple and systematic way of selecting these matrices.

As Q is increased, the penalty for deviating from the nominal trajectory is increased if R is unchanged. The resulting trajectory tends to be closer to the nominal trajectory, but at the expense of greater control requirements. When R is large, which means the cost of control is important, the states decrease slowly. As R decreases, the control becomes quite large.

The proper selection of weighting matrices calls for physical and engineering intuition and simulation experience with different trial runs. However, there are some techniques available for choosing the weighting matrices [9]. These are: (i) Heuristic Methods, (ii) Optimal Modal Control Methods, (iii) Asymptotic Optimal Root Loci Methods, and (iv) Dynamic Weighting Methods.

6.5.1 Heuristic Methods

One of the heuristic methods is the 'inverse-square method' based on rule-of-thumb [10]. This is still a widely used method for the selection of quadratic cost matrices. The basic idea is the normalization with respect to their expected maximum values. Also, in most practical applications, the matrices Q and R are selected to be diagonal so that specific components of the state and control perturbation vectors can be penalized individually. Thus,

$$Q = \begin{bmatrix} \dfrac{1}{\delta h_M^2} & 0 & 0 & 0 & 0 \\ 0 & \dfrac{1}{\delta v_M^2} & 0 & 0 & 0 \\ 0 & 0 & \dfrac{1}{\delta \gamma_M^2} & 0 & 0 \\ 0 & 0 & 0 & \dfrac{1}{\delta \phi_M^2} & 0 \\ 0 & 0 & 0 & 0 & \dfrac{1}{\delta \psi_M^2} \end{bmatrix} \quad ; \quad R = \begin{bmatrix} \dfrac{1}{\delta \eta_M^2} & 0 \\ 0 & \dfrac{1}{\delta \sigma_M^2} \end{bmatrix}$$

6.5.2 Optimal Modal Control Methods

The modal control approach is to determine a state feedback law u = -Fx, so that the closed-loop system matrix $D(= A\text{-}BF)$ has a prescribed set of eigenvalues [11]. Optimal modal control is based on the conventional pole-placement technique, but, instead of choosing the feedback matrix $F(= R^{-1}B^TP)$ directly, the weighting matrices of the performance index in linear quadratic regulator problem, are chosen to achieve the objective of placing the eigenvalues at the desired locations. The problem is to find the state weighting matrix Q for a given control weighting matrix R that corresponds to a set of prescribed eigenvalues. The method is based on transforming the given linear system into a diagonal (decoupled) system. The approach uses explicit formulas relating the elements of the state weighting matrix, the actual and the desired eigenvalues.

In the modal control approach, the choice of feedback matrix F is not unique, so that there may be a number of weighting matrices giving the same closed-loop eigenvalues. However, the method is easy for computer implementation. The details of the method are omitted and only the algorithm is given below for the case of real eigenvalues [11].

6.5.3 Algorithm

Step 1: Initialization (i=0)

Given the system matrices A, and B and the weighting matrices Q, and R, we obtain the optimal feedback matrix F_i using LQR theory.

Step 2: Loop

Obtain the system matrix $D_i = A - BF_i$

Step 3: Eigenvalues/eigenvectors

(i) Compute modal (eigenvector) matrix $M_i = [v_1,.....,v_n]$

(ii) Compute diagonal matrix $\Lambda_i = M_i^{-1} D_i M_i^T$

(iii) Compute matrix $H_i = M_i^{-1} BR^{-1}B^T M_i^{-T}$

Step 4: Update $i = i+1$

Step 5: Shifting of eigenvalues

The actual eigenvalue λ_{aj} is to be shifted to the desired position λ_{dj}. Compute,

$$\left[\tilde{q}_{jj}\right]_i = \frac{\lambda_{dj}^2 - \lambda_{aj}^2}{\left(h_{jj}\right)_{i-1}}$$

Step 6:

Form $\tilde{Q}_i = \text{diag}\left\{0,0,....,\left[\tilde{q}_{jj}\right]_i,0,....,0\right\}$

Step 7: Matrix Riccati equation

Using $\tilde{Q}_i$ from step 6, solve for $\tilde{P}_i$ from the matrix Riccati algebraic equation,

$$\tilde{P}_i \Lambda_{i-1} + \Lambda_{i-1} \tilde{P}_i - \tilde{P}_i BR^{-1}B^T \tilde{P}_i + \tilde{Q}_i = 0$$

Step 8: Feedback matrix

(i) Obtain the feedback matrix $\tilde{F}_i = R^{-1}B^T \tilde{P}_i$

(ii) Compute $\bar{F}_i = \tilde{F}_i M_{i-1}^{-1}$

(iii) Update $F_i = F_{i-1} + \bar{F}_i$

Step 9: Weighting matrix

(i) Compute $\dot{Q}_i = M_{i-1}^{-T} Q_i M_{i-1}^{-1}$

(ii) Update $Q_i = Q_{i-1} + \dot{Q}_i$

Step 10: If the required number of eigenvalues are not shifted, then change *j* and go to step 2.

In the present work, a combination of the inverse square method and modal control method is adapted in arriving at the required weighting matrix for state deviations.

6.6 Numerical Example

The nominal trajectory corresponds to the fuel-optimal condition of orbital transfer with plane change [7]. At a particular point of time, the set of nominal values used is

altitude, H = 44724.24 m

velocity, V = 9278.73 m/sec

flight path angle, γ = -1.092 deg

latitude, ϕ = 0.1954 deg

heading angle, ψ = 9.424 deg

lift coefficient, C_L = 0.3231

bank angle, σ = 75.28 deg

With this set of nominal values, the linearized model is obtained and used for LQ regulator. For a typical set of perturbations, the solutions are obtained and presented in Figs. 2 and 3. Figs. 2(a) through 2(e) show the state perturbations, while Figs. 3(a) and 3(b) present perturbed controls. The guidance scheme is tested at different points on the nominal trajectory and this found to be good.

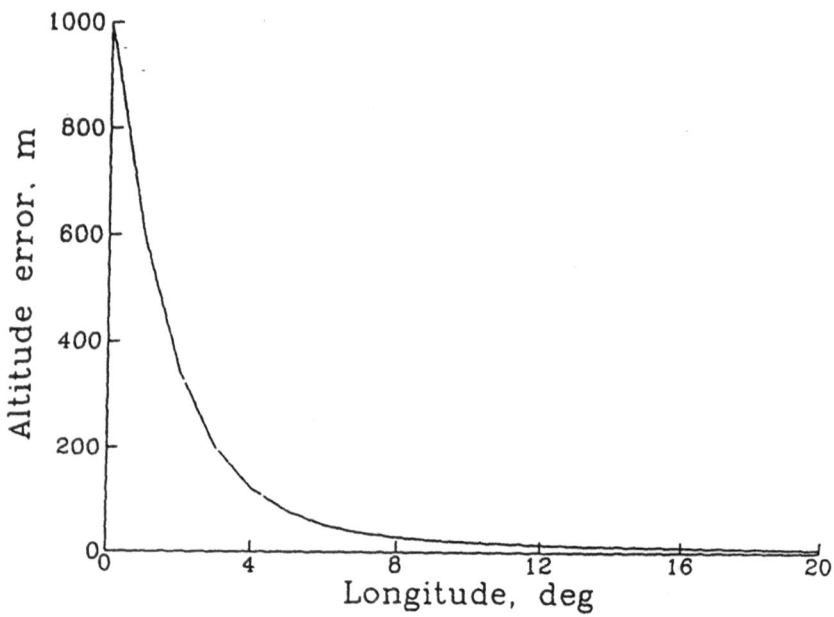

Fig. 2(a)   History of Altitude Error

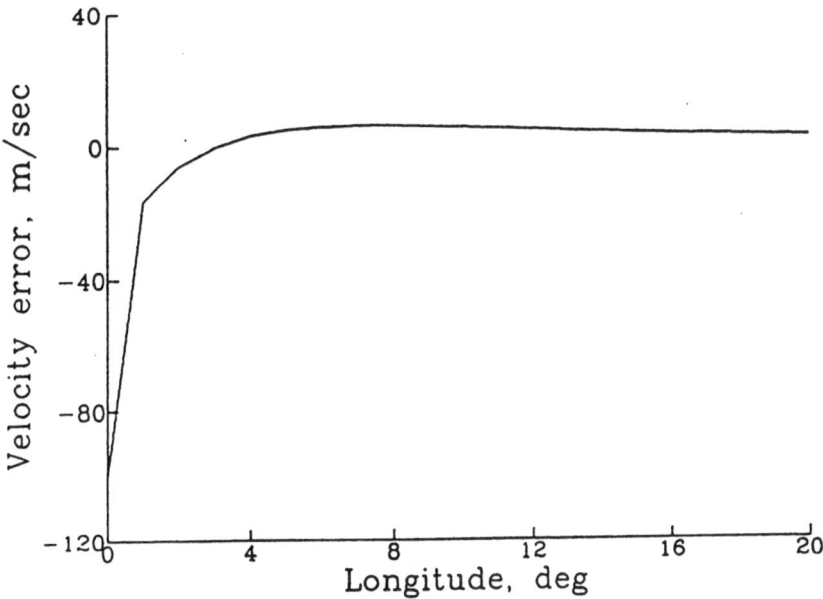

Fig. 2(b) History of Velocity Error

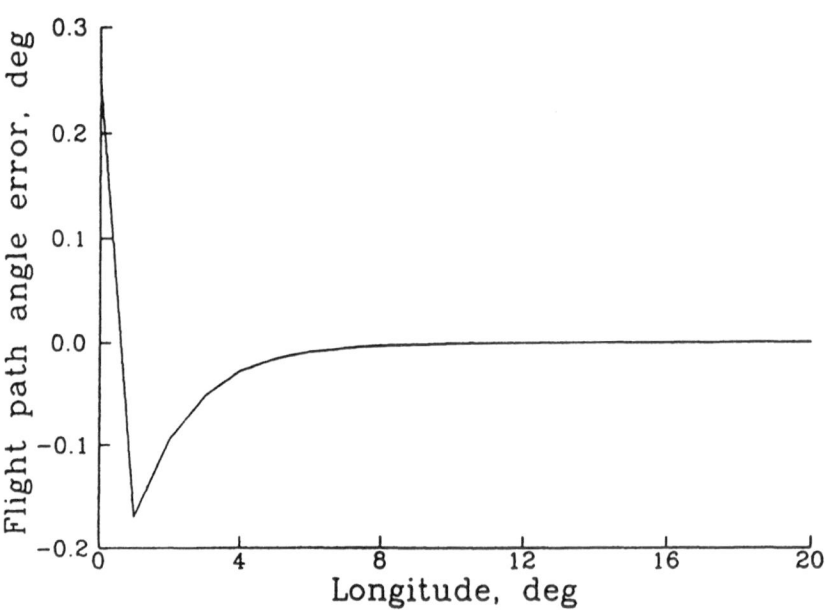

Fig. 2(c)   History of Flight Path Angle Error

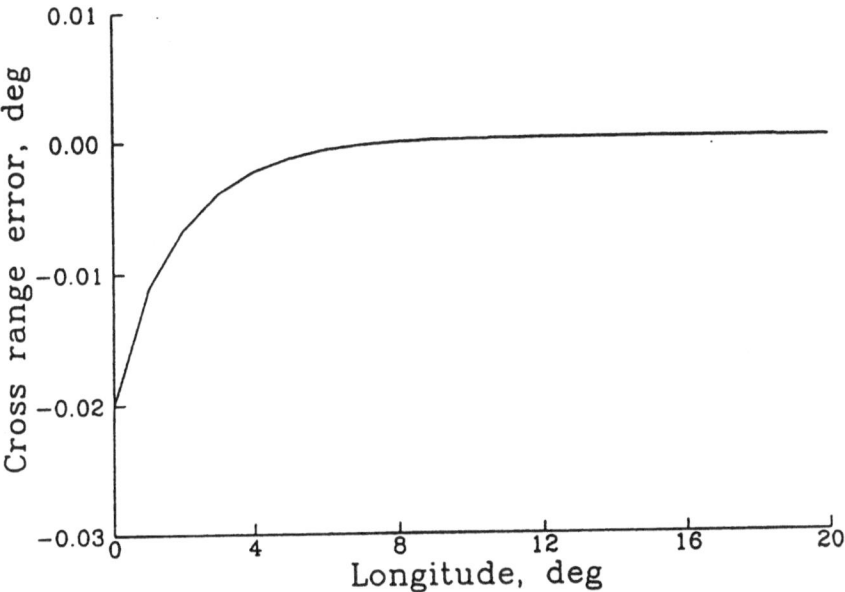

Fig. 2(d) History of Cross Range Error

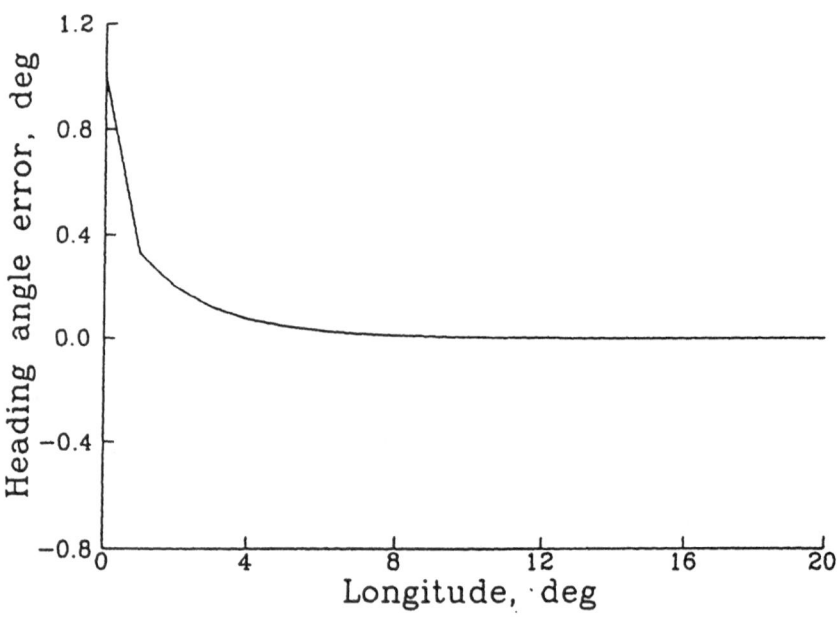

Fig. 2(e)   History of Heading Angle Error

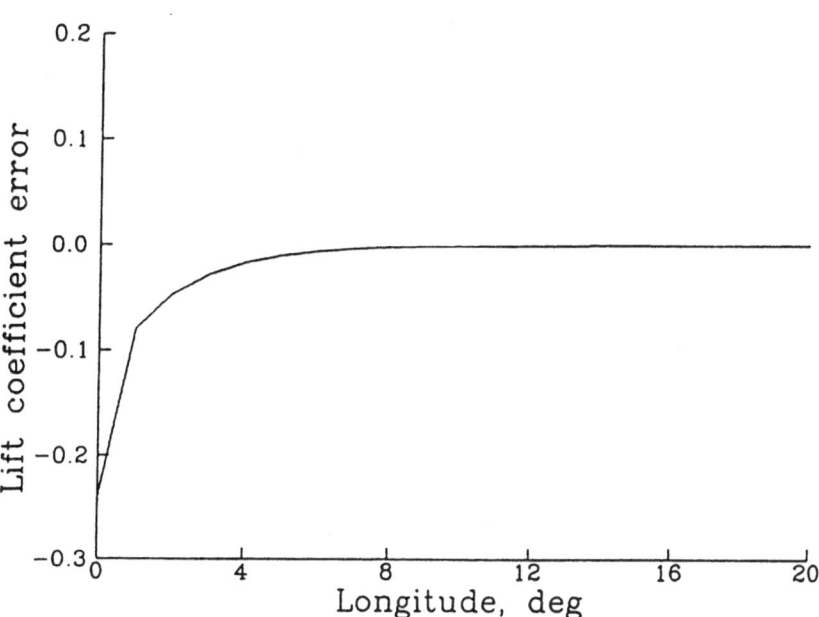

Fig. 3(a) History of Lift Coefficient Error

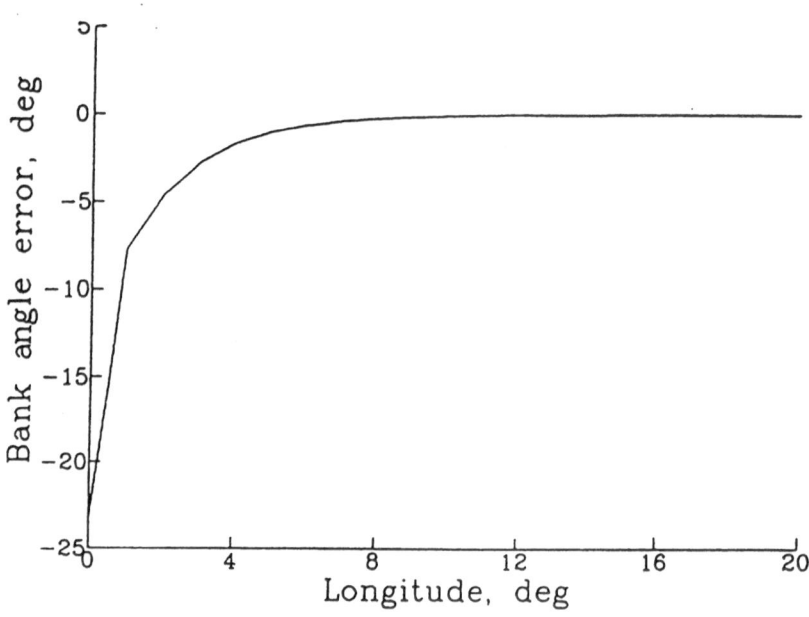

Fig. 3(b)   History of Bank Angle Error

6.7 Concluding Remarks

A neighboring optimal guidance scheme based on linear quadratic regulator has been designed for controlling an aeroassisted orbital transfer vehicle performing a plane change. The down range angle, instead of time, has been used as an independent variable. The problem has been formulated as a linear quadratic regulator with perturbed states consisting of altitude, velocity, flight path angle, latitude, and heading angle, and perturbed controls consisting of lift coefficient, and bank angle. The weighting matrices in the performance index have been chosen using a combination of heuristic approach and optimal modal control method. The results have been presented for fuel-optimal condition of an aeroassisted orbital transfer vehicle.

Nomenclature

$A_1 = C_{DO} S \rho_s H_a / 2m$

$A_2 = C_{LR} S \rho_s H_a / 2m$

$b = R_a / H_a$

C_D : drag coefficient

C_{DO}: zero-lift drag coefficient

C_L : lift coefficient

C_{LR}: lift coefficient for maximum lift-to-drag ratio

D : drag force

g : gravitational acceleration

H : altitude

h : normalized altitude

K : induced drag factor

L : lift force

m : vehicle mass

R : distance from Earth center to vehicle center of gravity

R_a : radius of atmospheric boundary

R_E : radius of Earth

S : aerodynamic reference area

t : time

V : velocity

v : normalized velocity

β : inverse atmospheric scale height

γ : flight path angle

ψ : heading angle

σ : bank angle

θ : down range angle
ϕ : cross range angle
δ : normalized density
μ : gravitational constant of Earth
η = normalized lift coefficient
ρ : density
ρ_s : density at the sea level
τ : normalized time

References

[1] G. D. Walberg, "A survey of aeroassisted orbital transfer," *J. Spacecraft*, 22, 3-18, 1985.

[2] M. A. De Virgilio, G. R. Wells, and E. E. Schiring, "Optimal guidance for aerodynamically controlled reentry vehicles," *AIAA Journal*, 12, 1331-1335, Oct. 1974.

[3] D. D. Sworder, "A feedback regulator for following a reference trajectory," *IEEE Trans. Aut. Control*, 22, 957-959, 1977.

[4] H. J. Pesch, "Real-time computation of feedback controls for constrained optimal control problems, part I: neighboring extremals," *Optimal Control: Applications and Methods*, 10, 129-145, 1989.

[5] D. G. Hull, and J.-Y. Lee, "Perturbation guidance for aerocruise with bounded control," *AIAA Guidance, Navigation, and Control Conference*, Boston, MA, Aug. 14-16, 1989.

[6] N. X. Vinh, *Optimal Trajectories in Atmospheric Flight*, Elsevier Scientific Publishing Co., Amsterdam, 1981.

[7] D. S. Naidu, "Fuel-optimal trajectories of aeroassisted orbital transfer with plane change," *AIAA Guidance, Navigation, and Control Conference*, Boston, MA, Aug.14-16, 1989.

[8] M. Athans, "The role and use of stochastic linear-quadratic-Gaussian problem in control system design," *IEEE Trans. Aut. Control*, 16, 529-552, Dec. 1971.

[9] M. A. Johnson, and M. J. Grimble, "Recent trends in linear optimal quadratic multivariable control system design," *IEE Proc*. Part D, 134, 53-71, 1987.

[10] A. E. Bryson, Jr., and Y.-C. Ho, *Applied Optimal Control*, Hemisphere

Publishing Corporation, Washington, D.C., 1975.

[11] O. A. Solheim, "Design of optimal control systems with prescribed eigenvalues," *Int. J. of Control*, 15, 143-160, 1972.

****

BIBLIOGRAPHY

The following bibliography arranged in alphabetical order refers to guidance and control strategies for aeroassisted orbital transfer. It is not intended to be exhaustive. However, the author has tried to make it as complete as possible. The author appreciates very much the help rendered by the following researchers in the field in compiling the bibliography. They are given in alphabetical order only:

M. D. Ardema, S. N. Balakrishnan, L. Beiner, R.D. Braun, A. J. Calise, C. J. Cerimele, M. Cupples, E. D. Dickmanns, J. H. Fay, W. G. Huber, D. G. Hull, H. Ikawa, T. S. Kuo, K. D. Mease, P. K. Menon, A. Miele, D. D. Moerder, J. C. Naftel, C. Park, D. W. Powell, D. B. Price, B. B. Roberts, J. L. Speyer, N.-X. Vinh, G. D. Walberg, W. H. Willcockson.

Aeroassist Flight Experiment Preliminary Design Document, NASA Marshall Space Flight Center, Huntsville, Alabama, May 27, 1986

Aeroassist Flight Experiment (AFE) Attitude Control System Design Handbook, NASA Marshall Space Flight Center, MSFC-DOC-1743, Huntsville, Alabama, May 1989.

Pioneering the Space Frontier, The Report of the National Commission on Space, Banton Books Inc., New York, 1986.

Andrews, D. G., and Caluori, V. A., "Optimization of aerobraked orbital transfer vehicles," AAIA 16th Thermophysics Conference, Palo Alto, CA, June 23-25, 1981.

Anthony, M. L., "Orbit adjustment maneuvers using very small tangential thrusts," 38th Congress of Intl. Astro. Federation, Brighton, UK, Oct. 10-17, 1987.

Austin, G., Bangsund, E., and Vinopal, T., "Designing the space transfer vehicle," Congress of the Intl. Astro. Federation, Malaga, Spain, Oct. 7-13, 1989.

Austin, R. E., Cruz, M. I., and French, J. R., "System design concepts and requirements for aeroassisted orbital transfer vehicles," AIAA 9th Atmospheric Flight Mechanics Conference, San Diego, CA, Aug. 9-11, 1982.

Avila, E., "Preliminary design trades of guidance, navigation, and   control concepts for low thrust orbit transfer, AIAA Guidance Navigation   and Control Conf., Hilton Head, SC, Aug. 10-12, 1992.

Balakrishnan, S., and Kamarsu, S., "Self tuning guidance applied to aeroassisted orbit transfer problems," AIAA Guidance, Navigation, and Control Conf., Hilton Head, SC, Aug. 10-12, 1992.

Bartholomen-Biggs, M. C., Dixon, L. W., Hersom, S. E., and Maany, Z. A., "The solution of some difficult problems in low-thrust interplanetary trajectory optimization, Optimal Control Applications Methods 9, 229-251, 1988.

Battani, R. H., "Space guidance evolution- a personal narrative," J. Guidance, Control, and Dynamics, 5, 97-110, March-April 1982.

Beiner, L., "A simplified three-dimensional altitude-constrained skip maneuver for maximum orbit plane change with given energy loss," Israel J. of Tech., 13, 82-88, 1975.

Beiner, L., "Maximum aerodynamic plane change with given energy loss by a three-dimensional altitude-constrained skip maneuver," Z. Flugwiss, 24, 286-291, 1976.

Beiner, L., "Fixed-range optimal re-entry maneuvers with bounded lift control," J. Flight Sci. & Space Research, 11, 161-166, 1987.

Beiner, L., "Optimal re-entry maneuvers with bounded lift control," J. Guidance, Control and Dynamics, 10, 321-329, 1987.

Bell, R. N., and Hankey, W. L., Jr., "Application of aerodynamic lift in accomplishing orbital plane change," Air Force Systems Command, Wright-Patterson Air Force Base, Ohio, ASD-TDR-63-693, Sept. 1963.

Bialla, P., and Henkley, M., "The space-based OTV and the establishment of the next launch site," 34th Congress of the Intl. Astro. Federation, Brighton, UK, Oct. 10-17, 1987.

Blakelock, J. H., Automatic Control of Aircraft and Missiles, John-Wiley & Sons, New York, 1965.

Blanchard, R. C., and Rutherford, J. F., "The shuttle orbiter high resolution accelerometer package experiment: preliminary flight results," AIAA 22nd Aerospace Sciences Meeting, Reno, NV, Jan. 9-12, 1984.

Blanchard, R. C., and Rutherford, J. F., "Shuttle orbiter high resolution accelerometer package (HIRAP): preliminary flight results," J. Spacecraft and Rockets, 22, 474-480, July-Aug. 1985.

Blissit, J. A. Jr., An adaptive guidance algorithm for an aerodynamically assisted orbital plane change maneuver, M. S. Thesis, Mass. Inst. of Tech, Cambridge, MA June 1988.

Boguslavskii, I. A., Filtering and Control, Optimization Software Inc., New York, 1986.

Bond, A., Martin, A. R., and Bond, R. A., "Concept studies for a laser powered orbital transfer vehicle," Acta Astronautica, 19, 73-86, 1989.

Bonner, M. M., Minimum fuel trajectories for the synergetic plane change maneuver, Ph. D. Thesis, Univ. of Michigan, Ann Arbor, MI, 1967.

Boussalis, D., Mease, K. D., and Wang, S. J., "Vehicle control during an atmospheric skip trajectory," AIAA Guidance and Control Conference, Monterery, CA, Aug., 1987.

Bradt, J. E., and Andrews, D. G., "Impact of upper atmospheric density distribution on aeroassisted orbital transfer vehicles," Proc. 35th Congress of the International Astronautical Federation, Polais de Beaulieu Lausanne, Switzerland, Oct. 8-13, 1984.

Bradt, J. E., Hardtla, J. W., and Cramer, E. J., "An adaptive guidance algorithm for aerospace vehicles," AIAA Guidance, Navigation and Control Conference, Snowmass, CO, Aug. 19-21, 1985.

Bradt, J. E., Jessick, M. V., and Hardtla, J. W., "Optimal guidance for future space applications," AIAA Guidance, Nav. & Control Conf., Monterery, CA, Aug. 17-18, 1987.

Brand, T. J., and Engel, A. G., "Aeroassist flight experiment guidance, navigation, and control," Annual Guidance and Control Conference, Keystone, CO, Feb. 1986.

Brand, T. J., Fuhry, D. P., and Shepperd, S. W., "An on board navigation system which fulfills Mars aerocapture guidance requirements," 27th AIAA Aerospace Meeting & Exhibit, Reno, Nevada, Jan. 1989.

Brandon, L. B., Reentry guidance and control for aeroassist flight experiment, Space Tech; Proc. of the Conf and Exposition, Anheim, CA, Sept. 1985.

Brauer, G. L., Cornick, D. E., and Stevenson, R., Capabilities and applications of the Program to Optimize Simulated Trajectories (POST), NASA CR-2770, Feb. 1977.

Braun, R. D., "The influence of interplanetary trajectory options on a chemically propelled manned Mars vehicle," J. of Astronautical Sciences, 38, 289-310, 1990.

Braun, R. D., and Powell, R. W., "Aerodynamic requirements of a manned Mars aerobraking transfer vehicle," AIAA Guidance, Navigation, and Control Conf., Portland, OR, Aug. 20-22, 1990.

Braun, R. D., and Powell, R. W., "A robust predictor-corrector guidance algorithm for use in high-entry aerobraking entry studies," AIAA 29th Aerospace Science Meeting and Exhibit, Reno, NV, 1991.

Braun, R. D., and Powell, R. W., "Predictor-corrector guidance algorithm for use in high-energy aerobraking system studies, J. Guidance, Control, and Dynamics, 15, 672-678, 1992.

Braun, R. D., Powell, R. W., and Hartung, L. C., "The effect of interplanetary trajectory options on a manned Mars aerobrake configuration," NASA Technical Paper, NASA TP-3019, Aug. 1990.

Braun, R. D., Powell, R. W., and Lyne, J. E., "Earth aerobraking strategies for manned return from Mars," J. Spacecraft and Rockets, 29, 297-3-4, May-June 1992.

Breakwell, J. V., and Rauch, H. E., "Optimum guidance for a low thrust interplanetary vehicle," AIAA Journal, 4, 693-704, April 1966.

Bryson, A. E. Jr., "New concepts in control theory, 1959-1984," J. Guidance, Control, and Dynamics, 8, 417-425, 1985.

Bryson, A. E., and Ho, Y. C., Applied Optimal Control, Hemisphere Publishing Company, New York, 1975.

Burns, R. E., "Forbidden tangential orbit transfers between intersecting Keplarian orbits," Acta Astronautica, 19, 649-656, 1989.

Busemann, A., and Vinh, N. X., "Optimum constrained disorbit by multiple impulses," J. Opt. Theory & Appl., 2, 40-64, 1968.

Busemann, A., Vinh, N. X., and Culp, R. D., "Solution of the exact equations for three-dimensional atmospheric entry using directly matched asymptotic expansions," NASA Contractor Report, CR-2643, March 1976.

Busemann, A., Vinh, A. X., and Kelley, G. F., "Optimum maneuvers of a skip
vehicle with bounded lift constraints," J. Opt. Theory & Appl., 3,
243-262, 1969.

Calise, A. J., "Singular perturbation analysis of the atmospheric orbital
plane change problem," The J. Astro. Sciences, 36, 35-43, Jan.-June,
1988.

Calise, A. J., and Melamed, N., "Optimal guidance of aeroassisted transfer
vehicles based on matched asymptotic expansions," AIAA Guidance,
Navigation, and Control Conference, New Orleans, LA, Aug. 12-14, 1991.

Carter, T. E., "Singular fuel-optimal space trajectories based on
linearization about a point in a circular orbit," J. Opt. Theory and
Appl., 54, 447-470, 1987.

Cerimele, C. J., and Gamble, J. D., "A simplified guidance algorithm for
lifting aeroassisted orbital transfer vehicles," AIAA 23rd Aerospace
Sciences Meeting, Reno, NE, Jan. 1985.

Cerimele, C. J., Skalecki, L. M., and Gamble, J. D., "Meteorological
accuracy requirements for aerobraking orbit transfer vehicles," AIAA
22nd Aerospace Sciences Meeting, Reno, NE, Jan. 1984.

Cervisi, R. T., "Analytic solution for a cruising plane change maneuver,"
AIAA Atmospheric Flight Mechanics Conference, Gatlingburg, TN, Aug.
15-17, 1983.

Chang, H., French, R., and Deaton, A., "Dynamic performance of aeroassist
flight experiment (AFE)," AIAA Aerospace Sciences Meeting and Exhibit,
Reno, NV, Jan. 7-10, 1991.

Charalambous, C., Hibey, J. L., and Naidu, D. S., "Neighboring optimal
guidance for aeroassisted orbital transfer with modeling and
measurement uncertainties," AIAA Aerospace Sciences Meeting and
Exhibit, Reno, NV, Jan.6-9, 1992.

Chatterjee, A. K., Optimal orbit transfer suitable for large space flexible
structures," The J. Astro. Sciences, 37, 261-280, 1989.

Cheatwood, F. M., DeJamette, F. R., and Hamilton, H. H., II, "Geometrical
description for a proposed aeroassist flight experiment vehicle," NASA
Technical Memorandum, NASA TM-87714, 1986.

Chern, J. S., Yang, C.-Y., and Sheen, J.-J., "Optimal lift and bank modulations for three-dimensional reentry trajectories with heat constraint," Acta Astronautica, 17, 303-309, 1988.

Chobotov, V., Ed., Orbital Mechanics, AIAA Educational Series, Waldorf, MD, 1991.

Clauss, J. S. Jr., and Yeatman, R. D., "Effect of heating restraints on aeroglide and aerocruise synergetic maneuver performance," J. Spacecraft & Rockets, 4, 1107-1109, 1967.

Cooper, L. P., and Scheer, D. D., "Status of advanced propulsion for space-based orbital transfer vehicle," Acta Astronautica, 17, 515-529, 1988.

Cruz, M. I., "The aerocapture vehicle mission design concept," Proc. on Conf. on Advanced Technology for Future Space Systems, NASA Langley Research Center, Hampton, VA, May 8-10, 1979.

Cruz, M. I., "Aerocapture vehicle mission design concepts for the inner and outer planets," AAS/AIAA Astrodynamics Specialist Conference, Province town, MA, June 25-27, 1979.

Cruz, M. I., "Trajectory optimization and closed-loop guidance of aeroassisted orbital transfer," AAS/AIAA Astrodynamics Conf., Lake Placid, New York, Aug. 1983.

Cruz, M. I., Armento, R. F., and Giles, W. H., "Aerocapture-a system design for planetary exploration," Congress of Intl. Astronautical Federation (IAF), Munich, FRG, Sept. 17-22, 1979.

Cruz, M. I., French, J. R., and Austin, R. E., "System design concepts and requirements for aeroassisted orbital transfer vehicles," AIAA 9th Atmospheric Flight Mechanics Conference, San Diego, CA, Aug. 9-11, 1982.

Cuadra, E., and Arthur, P. D., "Orbit plane change by external burning aerocruise," J. Spacecraft and Rockets, 3, 347-352, March 1966.

Cupples, M., Nordwall, J., LeDoux, S., and Woodcock, G., "Optimization of aero brake assisted descent trajectories at Mars atmosphere," AIAA Aerospace Sciences Meeting, Reno, NV, Jan. 7-10, 1991.

D'Amario, L. A., "Trajectory optimization software for planetary mission design," The J. Astro. Sciences, 37, 213-220, 1989.

D'Amario, L. A., Byrnes, D. V., and Stanford, R. H., "A new method for optimizing multiple-fly by trajectories," J. Guidance, Navigation, and Control, 4, 591-596, 1981.

D'Amario, L. A., Byrnes, D. V., and Stanford, R. H., "Interplanetary trajectory optimization with applications to Galilio," J. Guidance, Navigation, and Control, 5, 465-471, 1982.

Daum, A., "Solving the aeroassisted problem by using a sequential optimization approach," AIAA Guidance, Navigation, and Control Conf., Hilton Head, SC, Aug. 10-12, 1992.

Desautel, D., "Analytical characterization of AOTV perigee aerothermodynamic regime," AIAA 22nd Aerospace Science Meeting, Reno, NV, Jan. 9=12, 1984.

Da-yao, L., "Discussion on the optimal orbital transfer of multistage space vehicle," 40th Congress of Intl. Astro. Federation (IAF), Beijing, China, Oct. 8-14, 1989.

Davies, C. B., and Park, C., "Optimum configuration of high-lift aeromaneuvering orbit transfer vehicle in viscous flow," J. Spacecraft & Rockets, 25, 193-201, 1988.

Desautel, D., "Analytical characterization of AOTV perigee aerothermodynamic region," AIAA 22nd Aerospace Science Meeting, Reno, NV, Jan. 9-12, 1984.

Dickmanns, E. D., Gesleuerta drchung von satellitenbahnen durch eintauchen in die dichtere atmosphare, Ph. D. Thesis, RWTH, Aachen, Germany, 1969.

Dickmanns, E. D., "Optimal control for synergetic plane change," XXth Intl. Astronautical Congress, Mar del Plata, Argentina, Oct. 1969.

Dickmanns, E. D., "The effects of finite thrust and heating constraints on the synergetic plane change maneuver for a space shuttle orbiter-class vehicle," NASA TN D-7211, Oct. 1973.

Dickmanns, E. D., "Heating constrained synergetic plane change with finite thrust," XXVIth Intl. Astronautical Congress, Baku, USSR, Oct. 1973.

Dickmanns, E. D., "Comment on "minimum energy loss guidance for aeroassisted orbital plane change," J. Guidance, Control, and Dynamics, 9, 725, Nov.-Dec., 1986.

Durocher, C. L., and Darwin, C. R., "National space transportation and support study mission requirements and architecture studies," AIAA

Space System Technology Conf. San Diego, CA June 9-12, 1986.

Eckel, K. G., and Vinh, N. X., "Optimal switching conditions for minimum fuel fixed-time transfer between noncoplanar elliptical orbits," Acta Astronautica, 11, 621-631, 1984.

Eckstein, M. C., Shi, Y. Y., and Kevorkian, J., "Satellite motion for primary eccentricity and inclination around the smaller primary in the restricted three-body problem," Astron. Journal, 71, 248-264, May 1966.

Eisler, G. R., and Hull, D. G., "Optimal descending, hypersonic turn to heading," J. Guidance, Control, and Dynamics, 10, 255-261, May-June 1987.

Ess, R., "Atmospheric effects on Martian aerocapture," AIAA Guidance, Navigation, and Control Conf., Portland, OR, Aug. 20-22, 1990.

Eymar, P., Bonnal, C., Longstraff, R., and Salt, D., "ARIANE 5 transfer vehicle recent achievements," 40th Congress of Intl. Astro. Federation, Beijing, China, Oct. 8-14, 1989.

Fernandes, S. S., "Optimal low-thrust transfer between neighboring quasi-circular orbits around an oblate orbit," Acta Astronautica, 19, 933-938, 1989.

Findlay, J. T., and McConnell, J. G., "Atmospheres for aeroassisted orbital transfer vehicles based on Shuttle flight experience," Analytical Mechanics Associates, Inc., Hampton, VA, Sept. 1983.

Fitzgerald, S. M., and Ward, D. T., "Aeroassisted orbital transfer vehicle guidance performance in the presence of density dispersions," AIAA 26th Aerospace Science Meeting, Reno, Nevada, Jan. 11-14, 1988.

Florence, D., and Fisher, G., "System technology analysis of aeroassisted orbital transfer vehicles: moderate lift/drag,"AIAA Atmo. Mech. Conf., Gatlinburg, TN, Aug. 15-17, 1983.

Fouche, J. M., Nordwall, J. A., Rao, N. S., Tillotson, B. J., and Woodlock, G. R., "Neural networks based guidance for aerobraking applications," Boeing Aerospace & Electronics, Huntsville, Alabama, 1992.

Freeman, D. C., Powell, R. W., Naftel, J. C., and Wurster, K. E., "Definition of an entry research vehicle," J. Spacecraft, 24, 277-281, May-June 1987.

French, J. R., Jr., and Cruz, M. I., "Aerobraking and aerocapture for planetary missions," Astronautics & Aeronautics, 18, 48-55, 71, Feb. 1980.

French, J. R., Jr., and Uphoff, C. W., "Aerobraking for planetary missions," AAS Annual Meeting, Los Angels, CA, Oct.29-Nov.1, 1979.

Frostic, F., and Vinh, N. X., "Optimal aerodynamic control by matched asymptotic expansions," Acta Astronautica, 3, 319-332, 1976.

Fuhry, D. P., "A design study of on board navigation and guidance during aerocapture at Mars," Charles Stark Draper Lab., SDL-T-986, Cambridge, Massachusetts, May 1988.

Gamble, J. D., Cerimele, C. J., Moore, T. E., and Higgins, J., "Atmospheric guidance concepts for an aeroassisted flight experiment," The J. Astro. Sciences, 36, 45-71, Jan.-June, 1988.

Gamble, J. D., Cerimele, C. J., and Spartlin, K. M., "Aerobraking of a low L/D manned vehicle from GEO return to rendezvous with the space shuttle," AIAA Atmospheric Flight Mechanics Conference, Gatlingburg, TN, Aug. 15-17, 1983.

Gamble, J. D., Spratlin, K. M., and Skalecki, L. M., "Lateral directional requirements for a low L/D aeromaneuvering orbital transfer vehicle," AIAA Atmospheric Flight Conference, Seattle, WA, Aug. 21-23, 1984.

Garcia, F., Jr., and Fowler, W. T., "Thermal protection system weight minimization for the space shuttle through trajectory optimization," J. Spacecraft and Rockets, 11, 241-245, April 1974.

Geyling, F. T., and Westerman, H. R., Introduction to Orbital Mechanics, Addison-Wesley Publ. Company, Reading, MA, 1971.

Gilbert, E. G., Howe, R. M., Lu, P., and Vinh, N. X., "Optimal aeroassisted intercept trajectories at hyperbolic speeds," J. Guidance, Control, and Dynamics, 14, 123-131, 1991.

Giltner, J. M., "An optimal guidance law for the aeroassisted orbital transfer plane change maneuver," M. S. Thesis, Univ. of Texas, Austin, May 1984.

Gunn, C. R., "United States orbital transfer vehicle programs," 40th Congress of Intl. Astro. Federation (IAF), Mayaga, Spain, Oct. 7-12, 1989.

Gur, I., and Taratuta, A., "Multi pass aeroassisted transfer between coplanar elliptic orbits in the presence of atmospheric density variations I," AIAA Guidance, Navigation, and Control Conference, Portland, OR, Aug. 20-22, 1990.

Gurley, J. G., "Guidance for an aerocapture maneuver," J. Guidance, Control, and Dynamics, 16, 505-510, May-June 1993.

Haissig, C. M., Mease, K. D., and Vinh, N. X., "Minimum-fuel, power-limited transfers between coplanar elliptical orbits," 42nd Congress of Intl. Astro. Federation (IAF), Montreal, Canada, Oct. 7-11, 1991.

Hanson, J. M., Optimal Maneuvers of Orbital Transfer Vehicles, Ph.D. Thesis, Univ. of Michigan, Ann Arbor, MI, 1983.

Hanson, J. M., "Combining propulsive and aerodynamic maneuvers to achieve optimal orbital transfer," J. Guidance, Control, and Dynamics, 12, 732-738, 1989.

Hargraves, C., Johnson, F., Paris, S., and Rettie, I., "Numerical computation of optimal atmospheric trajectories," J. Guidance, and Control, 4, 406-414, Aug. 1981.

Heald, D. A., "Economic benefits of the OTV program," Acta Astronautica, 8, 1237-1249, 1981.

Henkey, M., "Secondary applications of expended orbital transfer systems," 40th Congress of Intl. Astro. Federation (IAF), Beijing, China, Oct. 8-14, 1989.

Higgins, J. P., "An aerobraking guidance concept for a low L/D AOTV," CSDL Memorandum, Charles Stark Draper Laboratory, Cambridge, MA, May 1984.

Hill, O., "An adaptive guidance logic for an aeroassisted orbital transfer vehicle," AAS/AIAA Astrodynamics Conference, Lake Placid, New York, Aug. 22-25, 1983.

Hill, O., "Use of aerodynamic braking to achieve orbit insertion about Mars," AAS/AIAA paper 85-418, August 1985.

Hohmann, D., "The attainability of heavenly bodies," NASA Technical Translation, F-44, 1960.

Hsu, F.-K., Kuo, T.-S., and Chern, J.-S., "Optimal aeroassisted orbital plane change with heating-rate constraint," J. Guidance, Control, and Dynamics, 13, 186-189, Jan. 1990.

Huber, W. G., "Orbital maneuvering vehicle: a new capability," Acta Astronautica, 18, 13-24, 1988.

Huber W. G., and Finnel III, W., "Orbital maneuvering vehicle, guidance, navigation and control," Proc. Annual Rocky Mountain Guidance and Control Conf., Keystone, CO, Feb. 1985.

Huber, W. G., Stephenson, A. G., and Baker, R., "Orbital maneuvering vehicle: a new capability," 34th Congress of the Intl. Astro. Federation, Brighton, UK, Oct. 10-17, 1987.

Hull, D. G., "New analytical results for AOTV guidance," AIAA 12th Atmospheric Flight Mechanics Conf., Snowmass, CO, Aug. 19-21, 1985.

Hull, D. G., Giltner, J. M., Speyer, J. L., and Maper, J., "Minimum energy loss guidance for aeroassisted orbital plane change," J. Guidance, Control and Dynamics, 8, 487-493, July-Aug. 1985.

Hull, D. G., and Lee, J.-Y., "Perturbation guidance for aerocruise with bounded control,"AAIA Guidance, Navigation, and Control Conf., Boston, MA, Aug. 14-16, 1989.

Hull, D. G., McClendon, J. R., and Spyer, J. L., "Aeroassisted orbital plane change using elliptic drag polar," The J. Astro. Sciences, 36, 73-87, Jan.-June, 1988.

Hull, D. G., McClendon, J. R., and Spyer, J. L., "Improved aeroassisted plane change using successive approximations," The J. Astro. Sciences, 36, 89-101, Jan.-June, 1988.

Hull, D. G., and Speyer, J. L., "Optimal reentry and plane change trajectories," The J. Astronautical Sciences, 30, 117-130, April-June 1982.

Ikawa, H., "A methodology for aerodecelerating entry trajectory analysis," AIAA Atmospheric Flight Mechanics Conference, Gatlingburg, TN, Aug. 15-17, 1983.

Ikawa, H., "Effect of rotating Earth for analysis of aeroassisted orbital transfer vehicles," J. Guidance, Control, and Dynamics, 11, 47-52, 1988.

Iwaka, H., "Endoatmospheric thrust termination condition to achieve a low Earth orbit," J. Spacecraft and Rockets, 29, 360-365, May-June 1992.

Iwaka, H., and Rudiger, T. F., "Synergetic maneuvering of winged spacecraft orbital plane change," J. Spacecraft and Rockets, 19, 513-520, Nov.-Dec. 1982.

Jezewski, D. T., "Optimal impulsive maneuvers and aerodynamic braking," Optimal Control: Applications & Methods, 6, 1-11, Jan.-March, 1985.

Johannesen, J. F., Optimal Aeroassisted Orbital Transfer involving Elliptical Orbits, Ph.D. Thesis, Univ. of Michigan, Ann Arbor, MI, 1985.

Johannesen, J. R., Vinh, N. X., and Mease, K. D., "Effect of maximum lift to drag ratio on optimal aeroassisted plane change," 12th Atm. Flt. Mech. Conf., Snowmass, CO, Aug. 19-20, 1985.

Jones, J. J., "The rationale for an aeroassist flight experiment," AIAA Thermophysics Conf., Honolulu, Hawaii, June 8-10, 1987.

Joosten, B. K., and Pierson, B. L., "Minimum-fuel aerodynamic orbital plane change maneuvers," AIAA 19th Aerospace Sciences Meeting, St. Louis, MO, Jan. 12-15, 1981.

Kaplan, M. H., Modern Spacecraft Dynamics and Control, John Wiley & Sons, New York, 1976.

Kechichian, J. A., Cruz, M. I., and Rinderle, E. A., "Optimization and closed-loop guidance of drag-modulated aeroassisted orbital transfer," AIAA Atmospheric Flight Mechanics Conference, Gatlingburg, TN, Aug. 15-17, 1983.

Kechichian, J. A., Vinh, N. X., Rinderdale, E. A., and Cruz, M. I., "Optimal non-lifting aeroassisted orbital transfer between coplanar circular orbits using drag-modulations, " Jet Propulsion Laboratory Document 314-280, August 1982.

Kelley, J. T., McKinnis, J. A., and Kofal, A. E., "Evolution of strategy for a family of space transfer vehicle (STV)," 42nd Congress of Intl. Astro. Federation (IAF), Montreal, Canada, Oct. 7-11, 1991.

Kerridge, S. J., "Aerobraking mission design-mission domain and mass performance," AAS/AIAA Astrodynamics Specialists Conference, Lake Tahoe, NV, Aug. 3-5, 1981.

Koelle, D. E., and Obersteiner, M., "Orbital transfer systems for Lunar missions," 42nd Congress of Intl. Astro. Federation (IAF), Montreal, Canada, Oct. 7-11, 1991.

Kreigsman, B. A., Richards, R. T., and Brand, T. J., Guidance and navigation system studies for entry research vehicle, CSDL-P-28 64, Cambridge, MA, April 1986.

Lawden, D. F., Optimal Space Trajectories for Space Navigation, Butterworth, London, 1963.

Lawden, D. F., "Optimal transfers between coplanar elliptic transfers," J. Guidance, Control, and Dynamics, 15, 788-791, 1992.

Lawden, D. F., "Time-closed optimal transfer by two impulses between coplanar elliptical orbits," J. Guidance, Control, and Dynamics, 16, 585-587, May-June 1993.

Lee, J. Y., Maximum Orbit Plane Change with Heat-Transfer Rate Considerations, Ph.D. Thesis, Univ. of Texas at Arlington, TX, May 1988.

Lee, B. S., and Grantham, W. L., "Aeroassisted orbital maneuvering using Lyapunov optimal feedback control," J. Guidance, Control, and Dynamics, 12, 237-242, 1989.

Lee, J. Y., and Hull, D. G., "Maximum orbital plane change with heat-transfer-rate considerations," J. Guidance, Control, and Dynamics, 13, 492-497, 1990.

Lettis, W. R., Jr., and Pelekanos, A., "Aeroassisted orbital transfer mission evaluation," AIAA 9th Atmospheric Flight Mechanics Conference, San Diego, CA, Aug. 9-11, 1982.

Lewis, M., and Weinstein, S., "The design of hypersonic wave riders for aero-assisted interplanetary trajectories," AIAA Aerospace Sciences Meeting and Exhibit, Reno, NV, Jan. 7-10, 1991.

London, H. S., "Change of satellite orbit plane by aerodynamic maneuvering," J. Aerospace Sciences, 29, 323-332, Mar. 1962.

London, H. S., "Comments on aerodynamic plane change," AIAA Journal, 1, 2414-2415, Oct. 1960.

Longuski, J. M., and Puig-Suari, J., "Hyperbolic aerocapture and elliptic orbit transfer with tethers," 42nd Congress of Intl. Astro. Federation (IAF), Montreal, Canada, Oct. 7-11, 1991.

Maper, J., Development an comparison of optimal guidance laws for aeroassisted orbital transfer, M. S. Thesis, Univ. of Texas, Austin,

May 1984.

Marec, J. P., Optimal Space Trajectories, Elsevier Scientific Publishing Co., Amsterdam, 1979.

Marec, J.-P., et al., IAF Astrodynamics Committee, "Recent progress in astrodynamics," Acta Astronautica, 17, 1049-1057, 1988.

Markopoulos, N., and Mease, K. D., "Thrust law effects on the longitudinal stability of hypersonic cruise," AIAA Atmospheric Flight Mechanics Conference, Portland, OR, Aug. 1990.

Markopoulos, N., Mease, K. D., and Vinh, N. X., "Thrust law effects on the long period modes of aerospace craft," AIAA Atmospheric Flight Mechanics Conference, Boston, MA, Aug. 1989.

Martin, A. R., Bond, A., and Bond, R. A., "Concept studies for a laser powered orbital transfer vehicle," 34th Congress of Intl. Astro. Federation (IAF), Brighton, UK, Oct. 10-17, 1987.

Maslen, S. H., "Synergetic turns with variable dynamics," J. Spacecraft and Rockets, 4, 1475-1482, Nov. 1967.

McEneaney, W. M., "Optimal aeroassisted guidance using Loh's term approximations," J. Guidance, Control, and Dynamics, 14, 368-376, 1991.

McEneaney, W. M., and Mease, K. D., "Error analysis for a Mars landing," AIAA/AAS Astrodynamics Conf., Minneapolis, MN, Aug. 15-17, 1988.

McEnearney, W. M., and Mease, K. D., "Error analysis for a guided Mars landing," J. Astro. Sciences, 39, Oct.-Dec. 1991.

Mease, K. D., "Optimization of aeroassisted orbital transfer: current status," The J. Astro. Sciences, 36, 7-33, Jan.-June, 1988.

Mease, K. D., Chatterjee, A. K., Purvis, C. R., and McCreary, F. A., "State estimation and parameter identification during aerocapture at Mars," AAS/AIAA Astrodynamics Specialist Conference, Kalispell, MT, Aug. 1987.

Mease, K. D., and Cruz, M. I., "Aeromaneuvering entry options at Mars," AAS/AIAA Astrodynamics Specialist Conference, Kalispell, MT, Aug. 1987.

Mease, K. D., Lee, J. Y., and Vinh, N. X., "Orbital changes during hypersonic aerocruise," The J. Astro. Sciences, 36, 103-137, Jan.-June 1988.

Mease K. D., and McGreary, F. A., "Atmospheric guidance law for planar skip trajectories," AIAA Atmospheric Flight Mechanics Conf., Snowmass, CO, Aug. 1985.

Mease, K. D., and Utashima, M., "Effect of heat rate constraint on minimum-fuel synergetic plane change," AAS/AIAA Spaceflight Mechanics Conference, Houston, TX, Feb. 1991.

Mease K. D., and Vinh, N. X., "Minimum-fuel aeroassisted coplanar orbit transfer using lift modulation," J. Guidance, Control and Dynamics, 8, 134-141, Jan.-Feb. 1985.

Mease, K. D., Vinh, N. X., and Kuo, S. H., "Optimal plane change during constant altitude hypersonic flight," AIAA Atmospheric Flight Mechanics Conference, Minneapolis, MN, Aug. 1988.

Mease, K. D., Vinh, N. X., and Kuo, S. H., "Optimal plane change during constant hypersonic plane change," J. Guidance, Control, and Dynamics, 14, 797-806, 1991.

Mease, K. D., Weidner, R. J., Kechichian, J., Wood, L. T., and Cruz, M. I., "Aerocapture: guidance, navigation and control," AIAA 9th Atmospheric Flight Mechanics Conference, San Diego, CA, Aug. 9-11, 1982.

Melamad, N., and Calise, A. J., "Evaluation of an optimal guidance algorithm for aeroassisted orbital transfer," AIAA Guidance, Navigation, and Control, Hilton Head, SC, Aug. 10-12, 1992.

Menees, G. P., "Aeroassisted vehicle design studies for a manned Mars mission," Intl. Astro. Federation (IAF), Brighton, UK, Oct. 1987.

Menees, G. P., Brown, K. G., Wilson, J. E., and Davies, C. B., "Aerothermodynamic heating and performance analysis of a high-lift aeromaneuvering AOTV concept," J. Spacecraft, 24, 198-204, May-June 1987.

Menees, G. P., Park, C., and Wilson, J. F., "Design and performance analysis of a conical aerobrake orbital transfer vehicle concept," in Thermal Design of Aeroassisted Orbital Transfer Vehicles, H. F. Nelson (Ed.), Vol. 96, AIAA Progress in Astronautics and Aeronautics, New York, 1985.

Menees, G. P., Wilson, J. F., and Adelman, H. G., "Synergetic plane-change capability of a conceptual aeromaneuvering orbit-transfer vehicle," AIAA Atm. Flt. Mech. Conf., Monterery, CA, Aug. 17-19, 1987.

Meyerson, R. E., and Cerimele, C. J., "Aeroassist vehicle requirements for a Mars rover/sample return mission," AIAA Aerospace Science Meeting, Reno, Nevada, Jan. 1988.

Miele, A., (Ed.), Special Issue on Hypervelocity Flight, J. Astro. Sciences, 36, 1-197, Jan.-June 1988.

Miele, A., and Basapur, V. K., "Approximate solutions to minimax optimal control problems for aeroassisted orbital transfer," Acta Astronautica, 12, 809-818, 1985.

Miele, A., Basapur, V. K., and Lee, W. Y., "Optimal trajectories for aeroassisted coplanar orbital transfer," J. Opt. Theory & Appl., 52, 1-24, Jan. 1987.

Miele, A., Basapur, V. K., and Lee, W. Y., "Optimal trajectories for aeroassisted noncoplanar orbital transfer," Acta Astronautica, 15, 399-412, June-July 1987.

Miele, A., Basapur, V. K., and Mease, K. D., "Nearly-grazing optimal trajectories for aeroassisted orbital transfer," The Journal of the Astronautical Sciences, 34, 3-18, Jan.-Feb., 1986.

Miele, A., and Lee, W. Y., "Optimal trajectories for the aeroassisted flight experiment," 40th Congress of Intl. Astro. Federation (IAF), Beijing, China, Oct. 8-14, 1989.

Miele, A., Lee, W. Y., and Mease, K. D., "Optimal trajectories for aeroassisted noncoplanar orbital transfer-part II," 38th Congress of Intl. Astro. Federation (IAF), Brighton, UK, Oct. 10-17, 1987.

Miele, A., Lee, W. Y., and Mease, K. D., "Nearly-razing optimal trajectories for noncoplanar aeroassisted orbital transfer," The J. Astro. Sciences, 36, 139-157, Jan.-June 1988.

Miele, A., Lee, W. Y., and Mease, K. D., "Optimal trajectories for LEO-to-LEO aeroassisted orbital transfer," Acta Astronautica, 18, 99-122, 1988.

Miele, A., and Venkataraman, P., "Optimal trajectories for aeroassisted orbital transfer," IAF Paper No. 83-331, 34th Congress of International Astronautical Federation, Budapest, Hungary, 1983.

Miele, A., and Venkataraman, P., "Optimal trajectories for aeroassisted orbital transfer," Acta Astronautica, 11, 423-433, 1984.

Miele, A., and Wang, T., "General solution for the optimal trajectory of an AFE-type spacecraft," 42nd Congress of the Intl. Astro. Federation (IAF), Montreal, Canada, Oct. 7-11, 1991.

Miele, A., and Wang, T., "Gamma guidance of trajectories for coplanar, aeroassisted orbital transfer," J. Guidance, Control, and Dynamics, 15, 255-262, Jan.-Feb. 1992.

Miele, A., Wang, T., and Basapur, V. K., "Primal and dual formulations of sequential-restoration algorithms for trajectory optimization problems," Acta Astronautica, 13, 491-505, 1988.

Miele, A., Wang, T., and Deaton, A. W., "Properties of the optimal trajectories for coplanar aeroassisted orbital transfer," J. Opt. Theory and Appl., 69, 1-30, 1991.

Miele, A., Wang, T., and Deaton, A. W., "Decomposition techniques and optimal trajectories for the aeroassisted flight experiment," J. Opt. Theory and Appl., 69, 201-234, 1991.

Miele, A., Wang, T., and Lee, W. Y., "Optimization and guidance of trajectories for coplanar aeroassisted orbital transfer," J. of Astro. Sciences, 38, 311-333, 1990.

Miele, A., Wang, T., Lee, W. Y., and Zhao, Z. G., "Optimal trajectories for the aeroassisted flight experiment, Acta Astronautica, 21, 735-747, 1990.

Miele, A., Wang, T., and Zhao, Z. G., "Optimal trajectories for aeroassist flight experiment," 40th Congress of Intl. Astro. Federation (IAF), Malago, Spain, Oct. 7-13, 1989.

Miele, A., Zhao, Z. G., and Lee, W. Y., "Optimal trajectories for the aeroassisted flight experiment, Part 1 equations of motion in an Earth-fixed system," Aero-Astronautics Report, No. 238, Rice University, Houston, TX 1989.

Miele, A., Zhao, Z. G., and Lee, W. Y., "Optimal trajectories for the aeroassisted flight experiment, Part 2 equations of motion in an inertial system," Aero-Astronautics Report, No. 239 Rice University, Houston, TX 1989.

Mishne, D., and Speyer, J. L., "Optimal control aeroassisted plane change maneuver using feedback expansions," AIAA Flight Mechanics Conference, Williamsburg, VA, Aug. 18-20, 1986.

Mishne, D., and Speyer, J. L., "A guidance law for the aeroassisted plane change maneuver in the presence of atmospheric uncertainties," Proc. 25th IEEE Conf. Decision & Control, Athens, Greece, Dec. 10-12, 1986.

Mulqueen, J. A., "Application of low lift-to-drag ratio aerobrakes using angle of attack variation and control," NASA Technical Memorandum, NASA TM-103544, Marshall Space Flight Center, Alabama, June 1991.

Mulqueen, J. A., and Coughlin, D., "Lunar mission aerobrake performance study," NASA Technical Memorandum, NASA TM-103562, Marshall Space Flight Center, Alabama, Dec. 1991.

Myers, J., "Titan IV/Centaur: an orbital transfer system for DOD payloads," 42nd Congress of the Intl. Astro. Federation(IAF), Montreal, Canada, Oct. 7-11, 1991.

Naidu, D. S., Singular Perturbation Methodology in Control Systems, Peter Peregrinus Ltd., Stevenage Herts, England, 1988.

Naidu, D. S., "Three-dimensional atmospheric entry problem using the method of matched asymptotic expansions," American Control Conference (ACC), Atlanta, GA, June 15-17, 1988.

Naidu, D. S., "Fuel-optimal trajectories of aeroassisted orbital transfer with plane change," AIAA Guidance, Navigation, and Control Conference, Boston, MA, Aug. 14-16, 1989.

Naidu, D. S., "Three-dimensional atmospheric entry problem using the method of matched asymptotic expansions," IEEE Trans. on Aerospace and Electronic Systems, 25, 660-667, Sept. 1989.

Naidu, D. S., Guidance and Control Strategies for Aerospace Vehicles, Final Research Report, Old Dominion University, Norfolk, VA, Aug. 1990.

Naidu, D. S., "Orbital plane change maneuver with aerocruise," AIAA Aerospace Sciences Meeting and Exhibit, Reno, NV, Jan. 7-10, 1991.

Naidu, D. S., "Fuel-optimal trajectories for aeroassisted orbital transfer with plane change," IEEE Trans. Aerospace and Electronic Systems, 27, 361-369, March 1991.

Naidu, D. S., "Neighboring optimal guidance for aeroassisted noncoplanar orbital transfer," AIAA Atmospheric Flight Mechanics Conference, New Orleans, LA, Aug. 12-14, 1991.

Naidu, D. S., "Neighboring optimal guidance for aeroassisted noncoplanar orbital transfer," Intl. J. Systems Science, 24, 563-575, March 1993.

Naidu, D. S., Aeroassisted orbital transfer: guidance and control strategies, Lecture Notes in Control and Information Sciences, Springer-Verlag, Berlin, 1993.

Naidu, D. S., Hibey, J. L., and Charalambous, C., "Optimal control of aeroassisted coplanar orbital transfer vehicles," 27th IEEE Conference on Decision and Control, Austin, TX, Dec. 7-9, 1988.

Naidu, D. S., Hibey, J. L., and Charalambous, C., "Fuel-optimal trajectories for aeroassisted coplanar orbital transfer problem," IEEE Trans. Aerospace and Electronic Systems, 26, 374-381, March 1990.

Naidu, D. S., Hibely, J. L., and Charalambous, C., "Neighboring optimal guidance for an aeroassisted orbital transfer vehicle in the presence of modeling uncertainties," AIAA Guidance, Navigation, and Control Conference, Portland, OR, Aug. 1990.

Naidu, D. S., Hibely, J. L., and Charalambous, C., "Neighboring optimal guidance for an aeroassisted orbital transfer vehicle in the presence of modeling uncertainties," IEEE Trans. on Aerospace and Electronic Systems, 29, Oct. 1993.

Naidu, D. S., and Li, L., "Orbital plane change maneuver with aerobraking for Mars mission," 31st IEEE Conference on Decision and Control, Tucson, AZ, Dec. 16-18, 1992.

Nelson, H. F., (Ed.), Thermal Design of Aeroassisted Orbital Transfer Vehicles, Progress in Astronautics, Vol., 96, Amer. Inst. of Aero. & Astro., Inc., New York, 1985.

Nyland, F. S., "The synegetic plane change for orbiting spacecraft," Rand Corporation Memorandum, Santa Monica, CA, Aug. 1962.

Nyland, F. S., "Hypersonic turning with constant bank angle control," 6th Symp. on Space Technology and Science, Tokyo, Japan, Nov. 1965.

Nyland, F. S., "Considerations of applying continuous thrust during synergetic plane changing," AAS/AIAA Astro. Specialists Conf., Jackson, WY, Sept. 3-5, 1968.

Oberle, H. J., "Numerical treatment of minimax optimal control problems with applications to the reentry flight path problem," J. Astro. Sciences, 36, 159-178, Jan.-June 1988.

Obersteiner, M. H., "Detailed analysis of 2/3-impulse Earth-Mars trajectories departing from a given Earth orbit," 40th Congress of the Intl. Astro. Federation (IAF), Beijing, China, Oct. 8-14, 1989.

Paine, J. P., " Use of lifting reentry vehicles for synergetic maneuvers," J. Spacecraft and Rockets, 4, 698-700, 1967.

Papadopoulos, P., Tauber, M., and Chang, I.-D., "Aerobraking in a dusty Martian atmosphere," AIAA/ASME 5th Joint Thermophysics and Heat Transfer Conference, Seattle, WA, June 18-20, 1990.

Parson, W. D., "Analytical solution of the synergetic turn," J. Spacecraft, 3, 1675-1678, Nov. 1966.

Pesch, H. J., "Neighboring optimum guidance of a space Shuttle orbiter-type vehicle, " J. Guidance, Control and Dynamics, 3, 386-391, 1980.

Popescv, M., "Functional analysis applications to the study of optimal transfer," 40th Congress of Intl. Astro. Federation (IAF), Beijing, China, Oct. 8-14, 1989.

Powell, R. W., and Braun, R. D., "A six-degree-of-freedom guidance and control analysis of Mars aerocapture," AIAA Aerospace Sciences Meeting and Exhibit, Reno, NV, Jan. 6-9, 1992.

Powell, R. W., Naftel, J. C., and Cunningham, M. J., "Performance evaluation of an aeroassisted research vehicle," AIAA 24th Aerospace Sciences Meeting, Reno, NV, Jan. 6-9, 1986.

Powell, R. W., Naftel, J. C., and Cunningham, M. J., "Performance evaluation of an entry research vehicle," J. Spacecraft, and Rockets, 24, 489-495, Nov.-Dec. 1987.

Powell, R. W., Stone, H. W., and Naftel, J. C., "Performance evaluation of the atmospheric phase of aeromaneuvering orbital transfer vehicles," AIAA 22nd Aerospace Sciences Meeting, Reno, NV, Jan. 9-12, 1984.

Rehder, J. J., "A linear feedback guidance law for an aeromaneuvering orbit-to-orbit shuttle," AAS Paper 75-065, July 1975.

Rehder, J. J., "Multiple pass trajectories for an aeroassisted orbital transfer vehicle," AIAA 22nd Aerospace Sciences Meeting, Reno, NV, Jan. 9-12, 1984.

Roberts, B. B., "Systems analysis and technology development for the NASA orbital transfer vehicle," AIAA 20th Thermophysics Conference,

Williamsburg, VA, Jan. 19-21, 1985.

Rossler, M., "Optimal aerodynamic-propulsive maneuvering for the optimal plane change of a space vehicle," J. Spacecraft, and Rockets, 4, 1678-1680, Dec. 1967.

Roy, A. E., Orbital Motion, Third Edition, Adam Hilger, Bristol, 1968.

Sarnecki, A. J., "Minimal orbital dynamics," Acta Astronautica, 17, 881-891, 1988.

Sawaya, G. Y., Feaster, T. A., Beck, H. D., and Steincamp, J. W., "Operations and support for the next generation space transportation systems," AIAA Space Systems Technology Conference, San Diego, CA, June 9-12, 1986.

Scot, C. D., Roberts, B. B., Nagy, K., Taylor, P., Gamble, J. D., Cerimele, C. J., Kroll, K. R., Li, C. P., and Ried, R. C., "Design study of an integrated aerobraking orbital transfer vehicle," NASA Tech. Memo., 58264, Johnson Space Center, Houston, TX, March 1985.

Scott, D. R., "The satellite transfer vehicle," 40th Congress of Intl. Astro. Federation (IAF), Beijing, China, Oct. 8-14, 1989.

Sengenes, P., Legenne, J., Goester, J. F., Alby, F., and Wagner, A., "Optimal strategies for HERMES autonomous guidance and control during orbital flight," 40th Congress of Intl. Astro. Federation (IAF), Beijing, China, Oct. 8-14, 1989.

Shi, Y.-Y., "Matched asymptotic solutions for optimum lift controlled atmospheric entry," AIAA Journal, 9, 2229-2238, Nov. 1971.

Shi, Y.-Y., and Eckstein, M. C., "Ascent or descent from satellite orbit law thrust," AIAA Journal, 4, 2203-2209, Dec. 1966.

Shi, Y.-Y., and Eckstein, M. C., "Uniformly valid asymptotic solution of nonplanar earth-to-moon trajectories in the restricted four-body problem," The Astronautical Journal, 72, 685-701, 1967.

Shi, Y.-Y., Nelson, R., and Young, D., "The application of nonlinear programming and collocation to optimal aeroassisted orbital transfer trajectory," AIAA Aerospace Sciences Meeting and Exhibit, Reno, NV, Jan. 6-9, 1992.

Shi, Y.-Y., Pottsepp, L., "Asymptotic expansions of a hypervelocity atmospheric entry problem," AIAA Journal, 7, 353-355, Feb. 1969.

Shi, Y.-Y., Pottsepp, L., and Eckstein M. C., "A matched asymptotic solution for skipping entry into planetary atmosphere," AIAA Journal, 9, 736-738, Apr. 1971.

Shipley, B. W., and Ward, D. T., "Control algorithm for aerobraking in the Martian atmosphere," AAS/AIAA Spaceflight Mechanics Conference, Houston, TX, Feb. 1991.

Simeonov, A. K., "Optimal flight conditions for aerodynamic maneuver of spacecraft," 34th Congress of Intl. Astro. Federation (IAF), Brighton, UK, Oct. 10-17, 1987.

Smith, W. L., "Evolution of servicing of orbital and expeditionary systems," Acta Astronautica, 19, 93-98, Jan. 1989.

Spyer, J. L., and Crues, E. Z., "An approximate atmospheric guidance law for aeroassisted plane change maneuvers," AIAA Guidance, Navigation, and Control Conf., Minneapolis, MN, Aug. 15-17, 1988.

Speyer, J. L., and Womble, M. E., "Approximate optimal atmospheric entry trajectories," J. Spacecraft and Rockets, 8, 1120-1125, Nov. 1971.

Spriesterbach, T., and Ross, M., "Effect of heating-rate on the fuel efficiency of aerocruise and aerobang maneuvers," AIAA Guidance, Navigation, and Control Conf., Hilton Head, SC, Aug. 10-12, 1992.

Striepe, S. A., Braun, R. D., and Powell, R. W., "Interplanetary trajectory optimization of Mars aerobrake missions with constrained atmospheric entry velocities," AAS Paper 91-421, Aug. 1991.

Takano, H., and Mease, K. D., "Nature of maximum heat rate segment of a minimum-fuel synergetic plane change," AIAA Guidance, Navigation, and Control Conf., Hilton Head, SC, Aug. 10-12, 1992.

Talay, T. A., White, N. H., and Naffel, J. C., "Impact of atmospheric uncertainties and viscous interaction effects on the performance of aeroassisted orbital transfer vehicles," AIAA 22nd Aerospace Science Meeting, Reno, NE, Jan. 9-12, 1984.

Tang, C. C., and Kwok, J. H., "Aerobraking techniques for planetary missions," AAS/AIAA Astrodynamics Specialists Conference, Lake Tahoe, NV, Aug. 3-5, 1981.

Taratuta, A., Mishne, D., and Gur, I., "Aeroassisted plane transfer between coplanar elliptical orbits during constant heating rate flight in the atmosphere," AIAA Guidance, Navigation and Control Conf., Hilton Head, SC, Aug. 10-12, 1992.

Tauber, M. E., and Yang, L., "Performance comparisons of maneuvering vehicles returning from orbit," AIAA Atmospheric Flight Mechanics Conference, Monterey, CA, Aug. 1987.

Theillier, F., and Salt, D., "ARIANE transfer vehicle (ATV): a generic spacecraft," 42nd Congress of the Intl. Astro. Federation (IAF), Montreal, Canada, Oct. 7-11, 1991.

Ting, L., and Brofman, S., "On take-off from circular orbit by small thrust," Z. Angen. Math. Mech., 44, 417-428, Oct. Nov., 1964.

Vinh, N. X., "General theory of optimal trajectory for rocket flight in a resisting medium," J. Opt. Theory and Appl., 11, 189-202, 1973.

Vinh, N. X., Optimal Trajectories in Atmospheric Flight, Elsevier Scientific Publishing Co., Amsterdam, 1981.

Vinh, N. X., "Optimal control of orbital transfer vehicles," AIAA Atmospheric Flight Mechanics Conf., Gatlinberg, TN, Aug. 15-17, 1983.

Vinh, N. X., Busemann, A., and Culp, R. D., Hypersonic and Planetary Entry Flight Mechanics, Univ. of Michigan Press, Ann Arbor, MI, 1980.

Vinh, N. X., and Hanson, J. M., "Optimal aeroassisted return for high Earth orbit with plane change," Acta Astronautica, 12, 11-25, Jan. 1985.

Vinh, N. X., and Johannesen, J. R., "Optimal aeroassisted transfer between coplanar elliptic transfers," Acta Astronautica, 13, 291-299, 1986.

Vinh, N. X., Johannesen, J. R., Mease K. D., and Hanson, J. M., "Explicit guidance of drag-modulated aeroassisted transfer between elliptical orbits," AIAA Guidance, Navigation and Control Conference, Seattle, WA, Aug. 20-22, 1984.

Vinh, N. X., Johannesen, J. R., Mease K. D., and Hanson, J. M., "Explicit guidance of drag-modulated aeroassisted transfer between elliptical orbits," J. Guidance, Control and Dynamics, 9, 274-280, May-June 1986.

Vinh, N. X., Kuo, S. H., and Marchal, C., "Optimal time-free nodal transfers between elliptic orbits," Acta Astronautica, 17, 875-880, 1988

Vinh, N. X., and Lu, P., "Chebyshev minimax problems for skip trajectories," The J. Astro. Sciences, 36, 179-197, Jan.-June 1988.

Vinh, N. X., and Marchal, C., "Analytical solution of a class of optimum orbit performance," J. Opt. Theory & Appl., 5, 178-196, 1970.

Vinopal, T., "Designing the space transfer vehicle (STV)-candidate concepts and technologies," 40th Congress of Intl. Astro. Federation (IAF), Beijing, China, Oct. 8-14, 1989.

Walberg, G. D., "Aeroassisted orbit transfer window opens on missions," Aeronautics & Astronautics, 12, 36-43, Nov. 1983.

Walberg, G. D., "A survey of aeroassisted orbit transfer," J. Spacecraft, 22, 3-18, Jan.-Feb. 1985.

Walberg, G. D., "A review of aerobraking for Mars missions," Intl. Astro. Federation (IAF), Paper 88-196, Oct. 1986.

Walberg, G. D., "Aerocapture for manned Mars mission-status and challenges," AIAA Atm. Flt. Mech. Conf., New Orleans, LA, Aug. 12-14, 1991.

Walberg, G. D., Siemers, P. M., and Jones, J. J., "The aeroassist flight experiment," 34th Congress of Intl. Astro. Federation, Brighton, UK, Oct. 10-17, 1987.

Wertz, J. R., Mullikin, T. L., and Brodsky, R. F., "Reducing the cost and risk of orbit transfer," J. Spacecraft and Rockets, 25, 75-80, Jan.-Feb.   1988.

Wickman, L. A., "Space-based servicing," Acta Astronautica, 15, 583-586, Aug. 1987.

Wiesel, W. E., Spaceflight Dynamics, McGraw Hill Book Company, New York, NY, 1989.

Wilhite, A. W., Arrington, J. P., and McCandless, R. S., "Performance aerodynamics of aeroassisted orbital transfer vehicles," AIAA 22nd Aerospace Sciences Meeting, Reno, NE, Jan. 9-12, 1984.

Willcockson, W. H., "Flight operation considerations for an aerobraked OTV," AIAA 23rd Aerospace Science Meeting, Reno, NV, Jan. 1985.

Willcockson, W. H., "Application of aeroassist to the Mars rover sample return mission," AAS/AIAA Intl. Symp. Goddard Space Center, Greenbelt, MD, April 24-27, 1989.

Willcockson, W. H., "OTV aeroassist with low L/D," Acta Astronautica, 17,

277-301, March 1988.

Willcockson, W. H., "Recent developments in aerocapture for the Mars rover sample return mission," 13th AAS Guidance and Control Conference, Keystone, CO, Feb. 1990.

Willcockson, W. H., "Mission benefits of aeroassist," 42nd Congress of the Intl. Astro. Federation (IAF), Montreal, Canada, Oct. 7-11, 1991.

Wingrove, R. C., "Survey of atmospheric reentry guidance and control methods," AIAA Journal, 1, 2019-2029, Sept. 1963.

Woodcock, G. R., and Sherwood, B., "Engineering aerobrakes for exploration missions," 40th Congress of Intl. Astro. Federation (IAF), Beijing, China, Oct. 8-14, 1989.

Woodcock, G. R., and Sherwood, B., "Engineering aerobrakes for exploration missions," Acta Astronautica, 21, 397-404, 1990.

Zhao, Y., and Garrard, W., "Feedback guidance laws for aeroassisted maneuvers," AIAA Guidance, Navigation, and Control Conf., Hilton Head, SC, Aug. 10-12, 1992.

****

INDEX

Lecture Notes in Control and Information Sciences

Edited by M. Thoma and A. Wyner

1989–1993 Published Titles:

Vol. 184: Duncan, T.E.; Pasik-Duncan, B. (Eds.)
Stochastic Theory and Adaptive Control. Proceedings of a Workshop held in Lawrence, Kansas, September 26-28, 1991.
500 pages. 1992 [3-540-55962-0]

Vol. 185: Curtain, R.F. (Ed.); Bensoussan, A.; Lions, J.L.(Honorary Eds.)
Analysis and Optimization of Systems: State and Frequency Domain Approaches for Infinite-Dimensional Systems. Proceedings of the 10th International Conference, Sophia-Antipolis, France, June 9-12, 1992.
648 pp. 1993 [3-540-56155-2]

Vol. 186: Sreenath, N.
Systems Representation of Global Climate Change Models. Foundation for a Systems Science Approach.
288 pp. 1993 [3-540-19824-5]

Vol. 187: Morecki, A.; Bianchi, G.; Jaworeck, K. (Eds.)
RoManSy 9: Proceedings of the Ninth CISM-IFToMM Symposium on Theory and Practice of Robots and Manipulators.
476 pp. 1993 [3-540-19834-2]

Vol. 188: Naidu, D. Subbaram
Aeroassisted Orbital Transfer: Guidance and Control Strategies.
192 pp. 1993 [3-540-19819-9]

Vol. 189: Ilchmann, Achim
Non-Identifier-Based High-Gain Adaptive Control
220 pp. 1993 [3-540-19845-8]

Vol. 190: Chatila, R; Hirzinger, G (Eds.)
Experimental Robotics II: The 2nd International Symposium, Toulouse, France, June 25-27 1991.
576 pp (approx.) 1993 [3-540-19851-2]